Introduction to
Industrial Polyethylene

Scrivener Publishing
3 Winter Street, Suite 3
Salem, MA 01970

Scrivener Publishing Collections Editors

James E. R. Couper	Richard Erdlac
Rafiq Islam	Pradip Khaladkar
Vitthal Kulkarni	Norman Lieberman
Peter Martin	W. Kent Muhlbauer
Andrew Y. C. Nee	S. A. Sherif
James G. Speight	

Publishers at Scrivener
Martin Scrivener (martin@scrivenerpublishing.com)
Phillip Carmical (pcarmical@scrivenerpublishing.com)

Introduction to Industrial Polyethylene

Properties, Catalysts, Processes

Dennis B. Malpass

Co-published by John Wiley & Sons, Inc. Hoboken, New Jersey, and Scrivener Publishing LCC, Salem, Massachusetts.
Published simultaneously in Canada.

For general information on our other products and services or for technical support, please contact our Customer Care Department within the United States at (800) 762-2974, outside the United States at (317) 572-3993 or fax (317) 572-4002.

Wiley also publishes its books in a variety of electronic formats. Some content that appears in print may not be available in electronic formats. For more information about Wiley products, visit our web site at www.wiley.com.

For more information about Scrivener products please visit www.scrivenerpublishing.com.

Cover design by Russell Richardson.

Library of Congress Cataloging-in-Publication Data:

ISBN 978-0-470-62598-9

10 9 8 7 6 5 4 3 2 1

Contents

Preface

Polyethylene is, by a wide margin, the largest volume synthetic polymer made by mankind. As of this writing, about 77 million metric tons are produced annually and the growth rate is expected to continue at about 5% per year into the foreseeable future. Within the three minutes or so it takes to read this preface, over 400 tons of polyethylene will have been manufactured. It is produced in various forms on 6 continents and its applications are ubiquitous in daily life, from the trash bag you placed on the curb this morning to Uncle Fred's artificial hip.

This book is primarily intended as an introductory text for chemists, engineers and students who wish to gain an understanding of the fundamentals of the commercially important polymers and copolymers of ethylene. The reader is assumed to have had a modicum of training in chemistry but little prior knowledge about polyethylene. I also intend it to be useful as a complement to courses on polymer chemistry. This book will answer essential questions such as:

- What are the types of polyethylene and how do they differ?
- What catalysts are used to produce polyethylene and how do they function?
- What is the role of cocatalysts in polyethylene production?
- What processes are used in the manufacture of polyethylene?
- What is the fate of polyethylene after its useful life is over?

Jargon used in industrial polyethylene technology can be bewildering to newcomers. This text will educate readers on terminology in common use in the industry and demystify the chemistry of catalysts and cocatalysts employed in the manufacture of polyethylene. I have employed several techniques to make the text user friendly. A thorough glossary is included in the appendix. The glossary not only provides definitions of acronyms and abbreviations, but also concisely defines terms commonly used in discussions of production and properties of polyethylene. An extensive index with liberal cross-referencing enables the reader to find a topic quickly.

Chapter 1 is used to review the history of polyethylene, to survey quintessential features and nomenclatures for this versatile polymer and to introduce transition metal catalysts (the most important catalysts for industrial polyethylene). Free radical polymerization of ethylene and organic peroxide initiators are discussed in Chapter 2. Also in Chapter 2, hazards of organic peroxides and high pressure processes are briefly addressed. Transition metal catalysts are essential to production of nearly three quarters of all polyethylene manufactured and are described in Chapters 3, 5 and 6. Metal alkyl cocatalysts used with transition metal catalysts and their potentially hazardous reactivity with air and water are reviewed in Chapter 4. Chapter 7 gives an overview of processes used in manufacture of polyethylene and contrasts the wide range of operating conditions characteristic of each process. Chapter 8 surveys downstream aspects of polyethylene (additives, rheology, environmental issues, etc.). However, topics in Chapter 8 are complex and extensive subjects unto themselves and detailed discussions are beyond the scope of an introductory text.

I must take this opportunity to express my appreciation to friends and associates who made constructive suggestions on the content of this book. Thanks to Drs. James C. Stevens and Rajen Patel (of The Dow Chemical Company in Freeport, TX) for their comments on the product descriptions and single-site catalysts. Dr. Roswell (Rick) E. King III (of Ciba, now part of BASF, in Tarrytown, NY) and Dr. Brian Goodall reviewed portions of the text and provided recommendations for improvement. Dr. Malcolm J. Kaus of ExxonMobil directed me to several outstanding literature discussions on catalyst and process technologies and kindly provided a reprint of a conference paper on the ExxonMobil high pressure process for polyethylene. Dr. James Strickler (of Albemarle Corporation in Baton Rouge, LA) helped with valuable suggestions about the chapter on metal alkyls. I am indebted to Drs. Balaji B. Singh and Clifford Lee of Chemical Marketing Resources, Inc. (Webster, TX) who shared information on markets and fabrication methods. Drs. Bill Beaulieu and Max McDaniel (of Chevron Phillips) reviewed information on "Phillips Catalysts" and instructed me on the intricacies of these mysterious catalysts. People mentioned above made suggestions that I found very helpful and tried to meld into the text. However, any residual errors are solely my responsibility. Finally, I would like to thank my publisher Martin Scrivener for the invitation to write this book and for his help in getting it published.

In closing, I would be remiss if I did not acknowledge my former colleagues at Texas Alkyls, Inc. (now Akzo Nobel) with whom I toiled for more than 30 years producing and marketing the metal alkyls that are so crucial to the polyolefins industry. The experiences and knowledge acquired during those years contributed mightily to the foundations for this book. However, my former coworkers at Texas Alkyls shall remain anonymous, for the list would be far too long.

I hope the reader will find the text informative on the fundamental aspects of industrial polyethylene.

Dennis B. Malpass
March 8, 2010

List of Tables

List of Figures

Chapter 8   Downstream Aspects of Polyethylene

1

Introduction to Polymers of Ethylene

1.1 Genesis of Polyethylene

Modern polyethylene has its origins in work by chemists at Imperial Chemical Industries beginning in 1933 (1). Eric Fawcett and Reginald Gibson were trying to condense ethylene with benzaldehyde at very high pressure and temperature (142 MPa* and 170 °C). They obtained a small amount of a residue that they concluded was polyethylene, but attempts to repeat the experiment minus benzaldehyde resulted in explosions. In late 1935, ICI chemist Michael Perrin succeeded in preparing a larger amount of polyethylene. Serendipitously, Perrin used ethylene containing traces of oxygen. Either oxygen itself or peroxides formed *in situ* initiated free radical polymerization of ethylene. In 1939, ICI began commercial production of high pressure polyethylene ("HPPE") now known as low density polyethylene (LDPE), and the product was used to insulate radar cable during World War II.

Work of other researchers portended the discovery of polyethylene. For example, in 1898, Hans von Pechman produced a composition he called "polymethylene" by decomposition of diazomethane. "Polymethylene" was also produced by other

---

* Please see glossary for definition of abbreviations, acronyms and terms.

1

chemistries, including the Fischer-Tropsch reaction. Most of these polymers had low molecular weights. In 1930, Marvel and Friedrich produced a low molecular weight polyethylene using lithium alkyls, but did not follow-up on this finding. Descriptions of early work on polyethylene have been provided by McMillan (1), Kiefer (2) and Seymour (3, 4).

Other noteworthy milestones in the evolution of industrial polyethylene include the following:

- In the early 1950s, transition metal catalysts that produce linear polyethylene were independently discovered by Hogan and Banks in the US and Ziegler in Germany.
- Gas phase processes, LLDPE and supported catalysts emerged in the late 1960s and 1970s.
- Kaminsky, Sinn and coworkers discovered in the late 1970s that an enormous increase in activity with metallocene single-site catalysts is realized when methylaluminoxane (discussed in Chapter 6) is used as cocatalyst.
- In the 1990s, polyethylenes produced with metallocene single site catalysts were commercialized and non-metallocene single-site catalysts were discovered by Brookhart and coworkers.

A timeline of notable 20th century polyethylene developments is provided in Figure 1.1.

This chapter introduces basic features of polyethylene, a product that touches everyday life in countless ways. However, polyethylene is not monolithic. The various types, their nomenclatures, and how they differ will be discussed. Key characteristics and classification methods will be briefly surveyed. An overview of transition metal catalysts has been included in this introductory chapter (see section 1.5) because these are the most important types of catalysts currently used in the manufacture of polyethylene. Additional details on transition metal catalysts will be addressed in subsequent chapters.

This chapter may be skipped by readers having an understanding of fundamental properties and nomenclatures of industrial polyethylene and a basic understanding of catalysts.

1.2 Basic Description of Polyethylene

Ethylene ($CH_2=CH_2$), the simplest olefin, may be polymerized (eq 1.1) through the action of initiators and catalysts. Initiators are most commonly

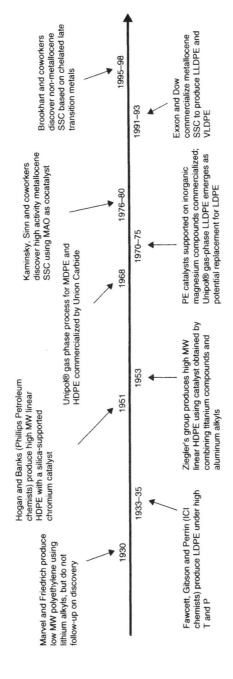

Figure 1.1 20th century milestones in polyethylene.

organic peroxides and are effective because they generate free radicals which polymerize ethylene via a chain reaction. Transition metal catalysts (primarily Ziegler-Natta and Phillips) are also widely employed in industry but produce polyethylene with different properties and by different mechanisms. Single-site catalysts also involve transition metal catalysts, but the quantity of polyethylene produced with single site catalysts at this writing is small (<4%). Initiators, transition metal catalysts and cocatalysts are discussed in Chapters 2–6.

Conditions for polymerization vary widely and polyethylene compositions, as noted above, also differ substantially in structure and properties. In eq 1.1, subscript n is termed the degree of polymerization (DP) and is greater than 1000 for most of the commercially available grades of polyethylene.

$$n\text{CH}_2 = \text{CH}_2 \xrightarrow{\text{catalyst}} \text{~}(\text{CH}_2\text{CH}_2)_n\text{~} \tag{1.1}$$

The polymer produced in eq 1.1 is known as polyethylene and, less commonly, as polymethylene, polyethene or polythene. (In the late 1960s, "polythene" became part of popular culture when the Beatles released "Polythene Pam.") Polyethylene is the IUPAC recommended name for homopolymer. As we shall see, however, many important ethylene-containing polymers are copolymers. Nomenclatures for various types of polyethylene are addressed in section 1.3. Though some have suggested that its name implies the presence of unsaturated carbon atoms, there are in fact few C=C bonds in polyethylene, usually less than 2 per thousand carbon atoms and these occur primarily as vinyl or vinylidene end groups.

Polyethylene is the least costly of the major synthetic polymers. It has excellent chemical resistance and can be processed in a variety of ways (blown film, pipe extrusion, blow molding, injection molding, etc.) into myriad shapes and devices. Fabrication methods will be briefly discussed in Chapter 8.

As removed from industrial-scale reactors under ambient conditions, polyethylene is typically a white powdery or granular solid. In most cases, the raw polymer is then melted and selected additives are introduced. (Additives are essential to improve stability and enhance properties of polyethylene. See Chapter 8.) The product is shaped into translucent pellets and supplied in this form to processors. Pelletization increases resin bulk density resulting in more efficient packing and lower shipping costs. It also lowers the possibility of dust explosions while handling.

Raw polyethylene resin is melted and shaped into pellets. This increases bulk density, improves handling characteristics and reduces shipping costs. Pellet size is typically ~3 mm (or ~0.1 in).

Polyethylene is a thermoplastic material. That is, it can be melted and shaped into a form which can then be subsequently remelted and shaped (recycled) into other forms. Polyethylene does not typically have a sharp melting point (T_m), but rather a melting range owing to differences in molecular weight, crystallinity (or amorphous content), chain branching, etc. Nevertheless, "melting points" between about 120 and 140 °C are cited in the literature. Because polyethylene is usually processed above 190 °C, where it is completely amorphous, melting ranges are less important than flow characteristics of the molten polymer. Molten polyethylene is a viscous fluid and is an example of what are termed "non-Newtonian fluids," that is, flow is not directly proportional to pressure applied (see section 8.3 of Chapter 8).

Polymerization of ethylene illustrated in eq 1.1 may be terminated by several pathways leading to different end groups. The type of end group depends upon several factors, such as polymerization conditions, catalyst and chain transfer agents used. Since end groups are primarily simple alkyl groups, polyethylene may be regarded as a mixture of high molecular weight alkanes.

Chain branching in low density versions of polyethylene is common. Extent and length of branching stem primarily from the mechanism of polymerization and incorporation of comonomers. Branching is classified as long chain branching (LCB) or short chain branching (SCB). By convention, SCB implies branches of 6 or fewer carbon atoms. LDPE contains extensive LCB and branches can contain hundreds of carbon atoms. Branches on branches are also common in LDPE.

This increases amorphous content and contributes to LDPE attributes, such as film clarity and ease of processing. As branching increases, density decreases. In LLDPE, incorporation of relatively large quantities of alpha olefin comonomers results in abundant SCB and lowering of density.

Ethylene may be copolymerized with a range of other vinylic compounds, such as 1-butene, 1-octene and vinyl acetate (VA). These are termed comonomers and are incorporated into the growing polymer. Comonomers that contain oxygenated groupings such as vinyl acetate are often referred to as "polar comonomers." Comonomer contents range from 0 to ~1 wt% for HDPE up to ~40 wt% for some grades of ethylene-vinyl acetate copolymer.

The range of suitable comonomers depends upon the nature of the catalyst or initiator. For example, Ziegler-Natta catalysts are poisoned by polar comonomers. Hence, commercial copolymers of ethylene and vinyl acetate are currently produced only with free radical initiators. However, some single site catalysts are tolerant of polar comonomers (see section 6.2.1).

When ethylene is copolymerized with substantial amounts (>25%) of propylene an elastomeric copolymer is produced, commonly known as ethylene-propylene rubber (EPR) or ethylene-propylene monomer (EPM) rubber. When a diene, such as dicyclopentadiene, is also included, a terpolymer known as ethylene-propylene-diene monomer (EPDM) rubber is obtained. EPR and EPDM are produced with single site and Ziegler-Natta catalysts and are important in the automotive and construction industries. However, EPR and EPDM are produced in much smaller quantities relative to polyethylene. Elastomers display vastly different properties than other versions of industrial polyethylene and are considered outside the purview of this text. EPR and EPDM will not be discussed further.

In copolymerizations of ethylene and α-olefins using Ziegler-Natta catalysts, ethylene is always the more reactive olefin. This causes compositional heterogeneity in the resultant copolymer. Composition distribution (CD) is the term applied to the uniformity (or lack thereof) of comonomer incorporation. For example, studies have shown that lower molecular weight fractions of LLDPE produced with Ziegler-Natta catalysts contain higher amounts of short chain branching, indicating nonuniform composition distribution. However, CD is highly uniform for ethylene copolymers made with single site catalysts.

Many grades of polyethylene are used in food packaging, *e.g.*, blow molded bottles for milk and blown film for wrapping meat and poultry. In the EU, the USA and other developed countries, the resin must satisfy governmental regulations for food contact. In the USA, the resin (including additives; see Chapter 8) must be compliant with FDA requirements for food contact, such as extractables and oxygen transmission rates. Catalyst residues are quite low in modern polyethylene and are considered to be part of the basic resin. Accordingly, catalyst residues are not subject to FDA regulations.

Polyethylene is available in a dizzying array of compositions, with different molecular weights, various comonomers, different microstructures, etc., predicated by selection of catalyst, polymerization conditions and other process options. Since 1933 when less than a gram was obtained unexpectedly from a laboratory experiment gone awry, polyethylene has grown to become the largest volume synthetically produced polymer, used today in megaton quantities in innumerable consumer applications. Recent analyses indicate global polyethylene production of about 77 million metric tons (~169 billion pounds) in 2008 (5).

1.3 Types and Nomenclature of Polyethylenes

Industrial polyethylenes are commonly classified and named using acronyms that incorporate resin density or molecular weight. IUPAC names are not typically used. In a few cases, copolymers are named using abbreviations for the comonomer employed. Nomenclature typically used for industrial polyethylenes will be discussed in this section. (Molecular weight will be discussed in section 1.4.)

Density is measured using density gradient columns and hydrostatic (displacement) methods. Density is directly related to crystalline content and in fact can be used to estimate % crystallinity in polyethylene.

The Society of the Plastics Industry (SPI), an industry trade association founded in 1937, identifies three main categories of polyethylene based on density:

- Low density: 0.910–0.925 g/cm^3
- Medium density: 0.926–0.940 g/cm^3
- High density: 0.941–0.965 g/cm^3

The American Society for Testing and Materials (ASTM) has also defined various types of polyethylene. An ASTM publication entitled "Standard Terminology Relating to Plastics" (ASTM D 883-00) provides the following classifications based upon density:

- high density polyethylene (HDPE): >0.941 g/cm^3
- linear medium density polyethylene (LMDPE): 0.926–0.940 g/cm^3

- medium density polyethylene (MDPE):   0.926–0.940 g/cm³
- linear low density polyethylene (LLDPE):   0.919–0.925 g/cm³
- low density polyethylene (LDPE)   0.910–0.925 g/cm³

While useful as starting points, SPI and ASTM classifications are not sufficient to describe the wide range of polyethylenes available in the industry. Classifications have been further subdivided to convey additional information, such as molecular weight or comonomer employed. Further, manufacturers use their own nomenclature and trade names. Clearly, the names used for various polyethylenes are somewhat arbitrary and subjective. The reader should not rigidly construe classifications and may encounter other nomenclatures. An overview of various classifications of polyethylene in common use in industry is provided below:

- *Very Low Density Polyethylene*: VLDPE, also called ultra low density polyethylene (ULDPE) by some manufacturers, is produced primarily with Ziegler-Natta catalysts using α-olefin comonomers. Density ranges from about 0.885 to 0.915 g/cm³. Selected grades of VLDPE produced with single site catalysts are known as polyolefin plastomers (POP) and polyolefin elastomers (POE) because products display both thermoplastic and elastomeric properties. Densities of POP fall in the VLDPE range, but densities of POE are in the range 0.855 to 0.885 g/cm³. Manufacturers have registered trademarks for POE and POP, such as Affinity®, Engage® and Exact®. A major application is in food packaging.

- *Low Density Polyethylene*: LDPE, the primeval polyethylene, is produced only by free radical polymerization of ethylene initiated by organic peroxides or other reagents that readily decompose into free radicals. Density is typically 0.915–0.930 g/cm³. LDPE is the most easily processed of major types of polyethylene and is often blended with linear low density polyethylene and high density polyethylene to improve processability. LDPE is highly branched and contains relatively high amorphous content which results in outstanding clarity in film for food packaging, a major application.

- *Copolymers of Ethylene with Polar Comonomers*:
  - *Ethylene-Vinyl Acetate Copolymer*: EVA is by far the most common copolymer of ethylene produced with a polar comonomer. EVA is produced by copolymerization of ethylene and vinyl acetate using free radical initiators, but cannot be produced with Ziegler-Natta or supported chromium catalysts. Vinyl acetate content ranges from about 8% to as high as 40%. Density is dependent upon amount of vinyl acetate incorporated, but is typically 0.93–0.96 g/cm³. When conducted under proper conditions (very high T), comonomer incorporation is random, resulting in uniform composition distribution. EVA contains high amorphous content which

results in outstanding clarity in film applications. EVA resins exhibit lower melting points than low density polyethylene and linear low density polyethylene and are useful for heat seal applications.

o *Ethylene-Vinyl Alcohol Copolymer*: EVOH is the name given to the specialty resin produced by alkaline hydrolysis of ethylene-vinyl acetate copolymer. It may be viewed as the product of copolymerization of ethylene with the hypothetical comonomer "vinyl alcohol." However, it is important to recognize that copolymerization of ethylene with vinyl alcohol is not possible. Vinyl alcohol exists in minute amounts (0.00006%) as the enol form in the keto-enol tautomerism of acetaldehyde (6). Density of EVOH is typically higher (0.96–1.20 g/cm³) than other types of polyethylene because of the high "vinyl alcohol" content. EVOH has excellent oxygen-barrier properties. A key application is laminated films for food packaging.

o *Ethylene-Acrylic Acid (EAA) and Ethylene-Methacrylic Acid (EMA) Copolymers*: EAA and EMA are produced by free radical copolymerization of ethylene with acrylic acid and methacrylic acid, respectively. An acrylate termonomer is incorporated in some grades to lower the glass transition temperature and to improve toughness. Like ethylene-vinyl acetate copolymer, EMA and EAA cannot be produced with Ziegler-Natta or supported chromium catalysts. Density is dependent upon amount of polar comonomer incorporated, but is typically 0.94–0.96 g/cm³. EAA and EMA copolymers have excellent adhesion to metals such as aluminum and are used in metal laminates. EAA and EMA are also used as precursors to ionomers. Ionomers are produced by reaction of a large portion (~90%) of the carboxylic acid moieties with metal bases, most often NaOH and $Zn(OH)_2$. Ionomers are rubbery solids at ambient temperatures, but become thermoplastic at higher temperatures. Though ionomers have low crystallinity, they exhibit excellent tensile strength and adhesion properties and form very tough films. A well known application for ionomers is their use as covers for durable (non-cut) golf balls. "Potassium ionomers" have recently been commercialized and are promoted for the antistatic properties they impart to blends (7).

• *Linear Low Density Polyethylene*: LLDPE is produced by copolymerization of ethylene with α-olefins using Ziegler-Natta, supported chromium or single site catalysts. LLDPE cannot be produced by free radical polymerization. Density is typically 0.915–0.930 g/cm³. Butene-1, hexene-1 and octene-1 are the most common comonomers, resulting in LLDPE with short chain branches of ethyl, *n*-butyl and *n*-hexyl groups, respectively. The quantity of comonomer incorporated varies depending upon the target resin. Typically, comonomer content is between about 2% and 4% (molar). Density decreases as greater amounts of comonomer are incorporated into the copolymer. (Figure 1.2 shows ethylene/alpha olefin copolymer densities as a function of comonomer content.) LLDPE has improved mechanical properties relative

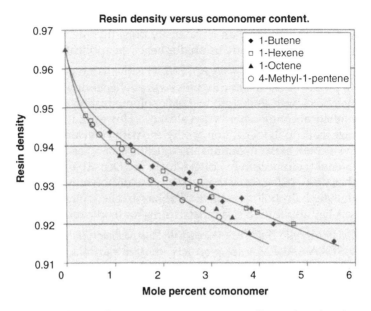

Figure 1.2 Density of LLDPE versus comonomer content (Reproduced with permission from *Kirk-Othmer Encyclopedia of Chemical Technology*, John Wiley and Sons, Inc., 6th edition, 2006).

to low density polyethylene and is used in blown and cast film applications, such as food and retail packaging. However, because it has lower amorphous content and heterogeneous composition distribution, LLDPE produces film that is not as clear as low density polyethylene made with free radical initiators.

- *Medium Density High Density Polyethylene*: MDPE (or MDHDPE) is produced by copolymerization of ethylene with α-olefins using Ziegler-Natta, supported chromium or single site catalysts. MDPE cannot be produced by free radical polymerization. MDPE has a linear structure similar to LLDPE, but comonomer content is lower. Density is typically 0.93–0.94 g/cm³. MDPE is used in geomembrane and pipe applications.

- *High Density Polyethylene*: HDPE is produced by polymerization of ethylene using Ziegler-Natta or supported chromium ("Phillips") catalysts. HDPE cannot be produced by free radical polymerization. Density is typically 0.94–0.97 g/cm³. Small amounts (<1%) of α-olefin comonomers are used in many of the commodity grades to introduce low concentrations of short chain branching, primarily to enhance processability but also to improve toughness and environmental stress crack resistance. HDPE has high modulus, yield and tensile properties relative to LLDPE and MDPE. However, because it has higher crystallinity, HDPE cannot match the clarity of LDPE or LLDPE film. HDPE is widely used in extruded pipe for potable water and gas distribution. Another important application is in blow molded packaging for household and industrial chemicals (HIC), such as bottles for bleach, shampoo, detergent, etc.

- *High Molecular Weight High Density Polyethylene*: HMW-HDPE (or HMWPE) is produced with Ziegler-Natta and supported chromium catalysts. HMW-HDPE cannot be produced by free radical polymerization. Molecular weight ranges from about 200,000 to 500,000 amu. Typically, density is in the range 0.94–0.96 g/cm³. HMW-HDPE grades are typically produced in a dual reactor configuration which results in bimodal molecular weight distribution with comonomer incorporated in the high molecular weight fractions. Key applications are pipe, grocery sacks and automotive fuel tanks.

- *Ultrahigh Molecular Weight Polyethylene*: UHMWPE is produced with Ziegler-Natta catalysts. Comonomers are not usually employed. Molecular weight ranges from about 3,000,000 to 7,000,000 amu. UHMWPE has a surprisingly low density (~0.94 g/cm³), most likely owing to crystalline defects and lamellar effects caused by the enormously long polymer chains. UHMWPE exhibits excellent impact strength and abrasion-resistance. Though difficult to process on standard equipment, UHMWPE may be compression-molded into prosthetic devices, such as artificial hips, or gel spun into very tough fibers used in bullet-proof vests worn by law enforcement personnel. UHMWPE is also used to make porous battery separator films.

- *Cyclic Olefin Copolymers*: COC are amorphous resins produced by copolymerization of ethylene with cyclic olefins, such as norbornene (8). COC are specialty resins with excellent optical properties and are produced with single site catalysts, but cannot be produced with conventional Ziegler-Natta catalysts. Relative to major types of polyethylene, COC are produced in very small quantities (<5 kT/y). On a molar basis, COC typically contain 40–70% ethylene. However, because of the relatively high molecular weight of comonomer, this amount corresponds to only 15–35% ethylene on a weight basis. For most commercial grades, density is 1.02–1.08 g/cm³. Key applications are in blister packaging for pharmaceuticals and anti-glare polarizing film (9).

- *Cross-linked Polyethylene*: XLPE (also PEX) is produced by cross linking polyethylene (mostly HDPE and MDPE) using free radicals generated by peroxides, ultraviolet or electron beam irradiation. A more intricate process involves grafting a vinylsilane compound, such as vinyltrimethoxysilane, onto polyethylene chains using free radicals. Subsequent moisture curing links adjacent chains through siloxide groupings. XLPE exhibits excellent environmental stress crack resistance (ESCR) and low creep. Used in piping for residential plumbing systems.

As previously noted, names given in the classifications above are those commonly used in the polyethylene industry, and these will be used throughout this text. IUPAC nomenclature is usually more complex. For example, IUPAC naming of a copolymer depends upon whether it is statistical, random, alternating, block, etc. If unspecified or unknown, the copolymer is named with the infix "co" included in its name. In the case of LLDPE made with ethylene and 1-butene, the IUPAC name would be poly(ethylene-*co*-1-butene). See Table 1.1 for additional examples of IUPAC names for common ethylene-containing polymers.

Microstructures of polyethylene depend upon type of catalyst, polymerization conditions, comonomers used, etc. Polyethylene and copolymers produced by free radical polymerization (LDPE, EVA, EAA, etc.) contain both short chain and long chain branching and higher amorphous content relative to LLDPE and HDPE. LLDPE and VLDPE contain extensive short chain branching owing to incorporation of α-olefin comonomers. HDPE contains little short chain branching because comonomers are used in low quantities, if at all. HDPE produced with Ziegler-Natta catalysts contains essentially no long chain branching. However, HDPE from Phillips catalysts contains very small amounts of long chain branching (10). Schematic representations of microstructures of several types of polyethylene are given in Figures 1.3–1.7. Short chain and long chain branching may be determined analytically by several techniques, for example, infrared (IR)

Table 1.1 IUPAC names of selected polymers and copolymers of ethylene.

Polymer Acronym	Comonomer	IUPAC Name
LDPE	none	polyethylene
VLDPE	butene-1	poly(ethylene-*co*-butene-1)
LLDPE	butene-1	poly(ethylene-*co*-butene-1)
LLDPE	hexene-1	poly(ethylene-*co*-hexene-1)
LLDPE	octene-1	poly(ethylene-*co*-octene-1)
LLDPE	4-methyl-pentene-1	poly(ethylene-*co*-4-methylpentene-1)
EVA	vinyl acetate	poly(ethylene-*co*-vinyl acetate)
EMA	methacrylic acid	poly(ethylene-*co*-methacrylic acid)
EVOH	vinyl alcohol*	poly(ethylene-*co*-vinyl alcohol)
HDPE	none**	polyethylene
COC	norbornene	poly(ethylene-*co*-norbornene)

* Hypothetical comonomer (see section 1.3)

** Small amount of α-olefin incorporated to improve polymer properties

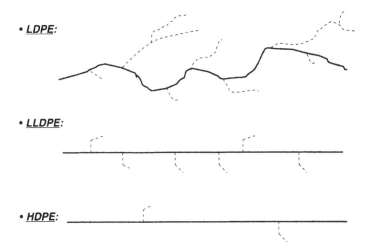

Figure 1.3 Schematic of microstructure of major types of polyethylene, where solid lines signify the "backbone" of the polymer and dashed lines represent SCB and LCB.

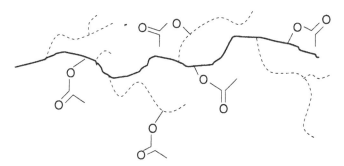

Figure 1.4 Schematic of microstructure of ethylene-vinyl acetate copolymer, where the solid line represents the "backbone" of the polymer and dashed lines signify SCB and LCB. Incorporation of VA results in pendant acetoxyl groups.

and nuclear magnetic resonance (NMR, both proton and carbon-13) spectroscopy and temperature rising elution fractionation (TREF).

Table 1.2 provides a summary of commonly used classifications in the polyethylene industry. A brief note is warranted here to conclude the survey of polyethylene classifications and nomenclature. In the early 1990s, several types of polyethylene manufactured with metallocene catalysts (a type of single site catalyst, see Chapter 6) were introduced to the market. To differentiate polyethylene produced with metallocenes from polyethylene manufactured using older conventional catalysts, metallocene grades are sometimes abbreviated mVLDPE, mLLDPE, etc.

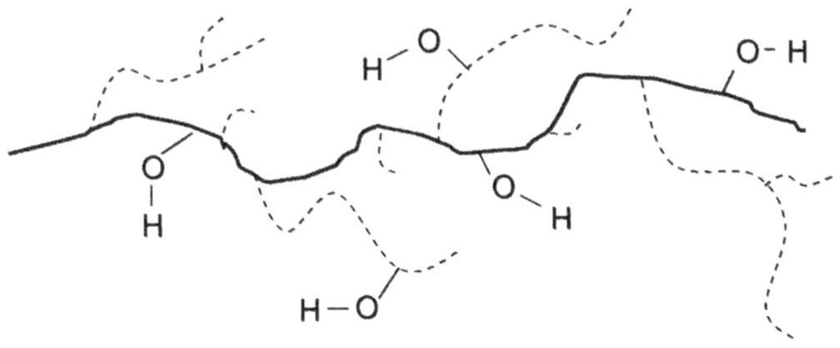

Figure 1.5 Schematic of microstructure of ethylene-"vinyl alcohol" copolymer, where the solid line signifies the "backbone" of the polymer and dashed lines represent SCB and LCB. Incorporation of VA results in pendant acetoxyl groups which are subsequently hydrolyzed to –OH groups.

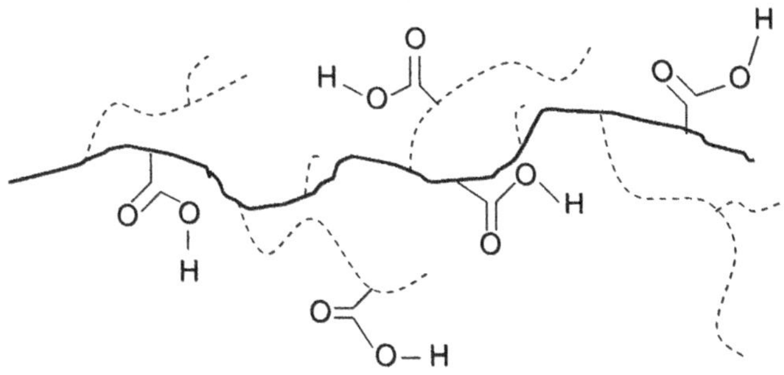

Figure 1.6 Schematic of microstructure of EAA, where the solid line represents the "backbone" of the polymer and dashed lines signify SCB and LCB. Incorporation of AA results in pendant carboxylic acid groups. EMA microstructure is similar, but includes a geminal methyl group.

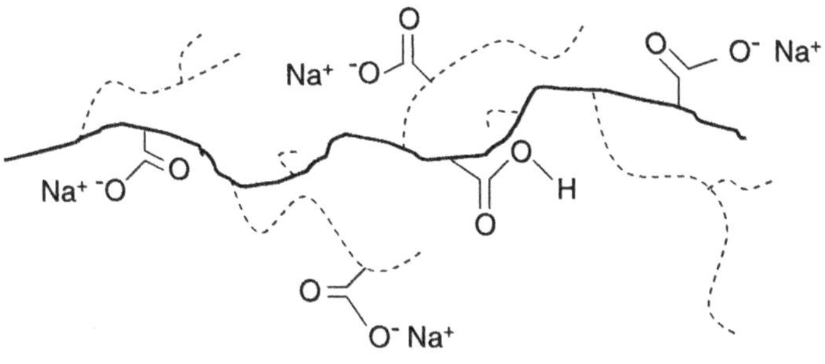

Figure 1.7 Schematic of microstructure of EAA ionomer, where the solid line signifies the "backbone" of the polymer and dashed lines represent SCB and LCB. Incorporation of AA results in pendant carboxylic acid groups which are ~90% converted to salts by reaction with a Bronsted base, such as NaOH. The microstructure of ionomer from EMA is similar, but includes a geminal methyl group.

Table 1.2 Classifications of selected polyethylenes.

Polymer	Approximate Density Range (g/cm³)	Typical Comonomer(s)	Catalysts (or Initiators) Used in Production
VLDPE[a]	0.88–0.91	α-olefins	ZN, SSC
LDPE	0.91–0.93	none	Organic Peroxides
EVA	0.93–0.97	vinyl acetate	Organic Peroxides
EAA/EMA[b]	0.94–0.96	AA, MA	Organic Peroxides
EVOH[c]	0.96–1.20	vinyl acetate	Organic Peroxides
LLDPE	0.91–0.93	α-olefins	ZN, supported Cr, SSC
MDPE	0.93–0.95	α-olefins	ZN, supported Cr
HDPE[d]	0.95–0.97	α-olefins[e]	ZN, supported Cr
UHMWPE	0.93–0.95	none	ZN
COC	1.02–1.08	norbornene	SSC

[a] Also known as ULDPE
[b] Precursors for production of ionomers
[c] Produced by saponification of EVA
[d] Includes XLPE
[e] Small amount of α-olefins often used to improve polymer properties

1.4 Molecular Weight of Polyethylenes

In addition to density, polyethylene manufacturers routinely supply data that correlate with molecular weight and molecular weight distribution. A measurement called the melt index (MI), also known as melt flow index (MFI), is determined by the weight of polyethylene extruded over 10 minutes at 190 °C through a standard die using a piston load of 2.16 kg. Reported in g/10min or dg/min, melt index is measured using an instrument called an extrusion plastometer according to ASTM D 1238-04c Condition 190/2.16, where the latter numbers refer to the temperature and the load in kg on the piston of the plastometer, respectively. Melt index is sometimes written "I_2" and is inversely proportional to molecular weight, *i.e.*, MI increases as molecular weight decreases. With the possible exception of density, melt index is the most widely cited property of industrial polyethylene.

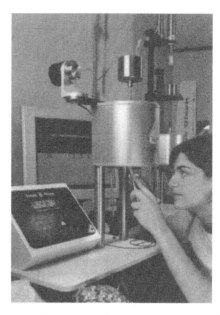

Melt index of polyethylene may be measured on an instrument called an extrusion plastometer. (Photo courtesy of Tinius Olsen.)

Another value is also measured on the plastometer at 190 °C, but under a load of 21.6 kg (ASTM D 1238-04c Condition 190/21.6). This is called the high load melt index (HLMI) and is also reported in g/10 min or dg/min. HLMI is often used for polyethylenes that have very high molecular weights. Because it may be difficult to measure the small amount of extrudate from the standard conditions for what are called "fractional" MI products (typically resins that have MIs <1), HLMI may be a more accurate measure of molten flow.

Dividing HLMI by MI affords the melt index ratio (MIR), a dimensionless number which gives an indication of breadth of molecular weight distribution. As MIR increases, MWD broadens.

$$MIR = HLMI \div MI \qquad (1.2)$$

In ASTM D 1238-04c, a term called the flow rate ratio (FRR) for polyethylene is defined as flow rate at Condition 190/10 divided by flow rate at Condition 190/2.16, abbreviated "I_{10}/I_2". Like MIR, the flow rate ratio is dimensionless and conveys information on molecular weight distribution.

The term "melt flow rate" (MFR) is sometimes applied (erroneously) to polyethylene. ASTM suggests "melt flow rate" be applied to other thermoplastics and "melt index" be reserved for polyethylene (see note 27 on p 10 of ASTM D 1238-04c). Melt flow rate is also determined under ASTM D 1238-04c, but different conditions are used in most instances. For example, MFR of polypropylene is determined at 230 °C and 2.16 kg.

MI and MIR measurements are inexpensive, relatively easy to conduct and are indicative of molecular weight and molecular weight distribution. Actual molecular weights may be determined using a variety of analytical methods, including gel permeation chromatography (GPC), viscometry, light scattering and colligative property measurements. (GPC is also called size exclusion chromatography or SEC.) However, these methods require more complex, sophisticated instruments, are more costly and difficult to perform and do not lend themselves to routine quality control procedures.

The number average molecular weight ($\overline{M}_n$) is calculated from the expression:

$$\overline{M}_n = \Sigma M_x N_x / \Sigma N_x \qquad (1.3)$$

where M_x is the molecular weight of the xth component and N_x is the number of moles of the xth component. Weight average molecular weight ($\overline{M}_w$) is calculated using the second order equation:

$$\overline{M}_w = \Sigma M_x^2 N_x / \Sigma M_x N_x \qquad (1.4)$$

The third order equation provides the "z-average molecular weight" and is calculated from the expression:

$$\overline{M}_z = \Sigma M_x^3 N_x / \Sigma M_x^2 N_x \qquad (1.5)$$

Higher order molecular weight averages may also be calculated, but are less important than $\overline{M}_w$, $\overline{M}_n$ and $\overline{M}_z$. For polydisperse polymers, such as polyethylene, the following relationship holds:

$$\overline{M}_z > \overline{M}_w > \overline{M}_n \qquad (1.6)$$

The ratio M_w/M_n is called the polydispersity index (also known as heterogeneity index and dispersity index) and is an indication of the broadness of molecular weight distribution. As polydispersity index increases, MWD broadens. If the polymer were a single macromolecule, the polydispersity would be 1.0 and the polymer would be said to be monodisperse.

For polyethylene produced with transition metal catalysts, molecular weight distribution is dictated largely by the catalyst employed. Polydispersities typically range from 2–3 for polyethylene made with single site catalysts, 4–6 for polymer produced with Ziegler-Natta catalysts and 8–20 for polyethylene made with supported chromium catalysts. These differences are illustrated

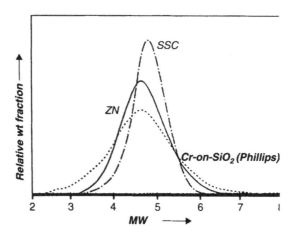

Figure 1.8 MWD of polyethylene from transition metal catalysts. (Reprinted with permission of John Wiley & Sons, Inc., *Kirk-Othmer Encyclopedia of Chemical Technology*, John Wiley and Sons, Inc., 6th edition, 2006).

schematically in Figure 1.8. Polyethylene with molecular weight distributions illustrated in Figure 1.8 are said to be unimodal.

In certain applications, such as blow molding, polyethylene with broad molecular weight distribution provides a better balance of properties. Higher molecular weight fractions impart mechanical strength while lower molecular weight fractions improve flow properties and make the polymer easier to process. The combination of good mechanicals and easier processing may also be realized if a resin with a bimodal molecular weight distribution can be produced.

Bimodal molecular weight distribution may be achieved by several techniques. The simplest method is post-reactor blending of polyethylene with different melt indices. Two other methods involve in-reactor production of polyethylene. One approach involves use of mixed catalyst systems that polymerize ethylene in different ways to produce polyethylene with different molecular weights. The latter requires that the catalysts are compatible. Another technique employs use of reactors in series operated under different conditions (see section 7.6 in Chapter 7). Figure 1.9 illustrates polyethylene with a bimodal molecular weight distribution produced with a single site catalyst system in a Unipol® gas-phase process.

Direct comparisons between melt index and molecular weight of polyethylene should be made with caution. Such comparisons are only appropriate when the polymers have similar histories (made using the same catalyst, by the same process, at near identical densities, etc.). An example of the relationship between melt index and molecular weight for a series of LLDPEs with similar histories is

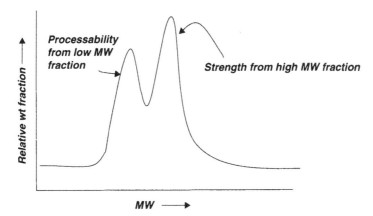

Figure 1.9 Bimodal MWD polyethylene produced with SSC in Unipol® gas phase process; P. J. Ferenz, 2nd Asian Petrochemicals Technology Conference, May 7–8, 2002, Seoul, Korea.

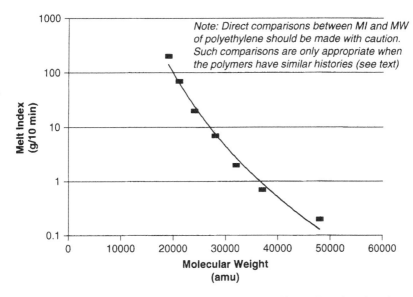

Figure 1.10 Molecular weight and melt index of LLDPE (d = 0.920 g/cc; data from H. Boenig, *Polyolefins: Structure and Properties*, Elsevier, p 80, 1966).

shown in Figure 1.10. While a valuable measurement, melt index provides little information regarding the shear sensitivity of the molten resin. Deformation under stress ("rheology," see section 8.3) of the resin is important to the performance of the polymer in its applications and requires additional testing.

1.5 Transition Metal Catalysts for Ethylene Polymerization

As previously mentioned, in addition to free radical initiation, ethylene may be polymerized by use of transition metal catalysts. To place the importance of these catalysts in proper perspective, one must recognize that transition metal catalysts were used to produce about 73% of the global industrial output of polyethylenes in 2008 or about 56 million tons (124 billion pounds).

In this section, we will introduce key characteristics of transition metal catalysts. Conditions used for polymerization with transition metal catalysts are much less severe than those needed for free radical processes. Transition metal catalysts used in production of polyethylene include Ziegler-Natta, supported chromium and single site catalysts, all to be discussed in more detail in subsequent chapters. The vast majority of these catalysts are produced using compounds of transition metals from Groups 4–6 of the Periodic Table. Ziegler-Natta catalysts (Chapter 3) are typically derived from inorganic titanium compounds. The most well-known and widely used supported chromium catalysts are the so-called "Phillips catalysts" (Chapter 5) but there are other industrially important chromium catalysts. Chromium catalysts must be supported on refractory oxides, most often silica, to be effective. Most commercial single-site catalysts (Chapter 6) involve Zr, Hf or Ti, but single site catalysts based on late transition metals, especially Pd, Fe and Ni, began to emerge in the mid-1990s.

All transition metal catalysts must fulfill several key criteria:

- Activity must be high enough to insure economic viability and that catalyst residues are sufficiently low in the final polymer to obviate post-reactor treatment. As a rule of thumb, this requires that catalyst activity exceed 150,000 lb of polyethylene per lb of transition metal.

- The catalyst must be capable of providing a range of polymer molecular weights:

  o For Ziegler-Natta and single-site catalysts, molecular weight is controlled primarily by use of hydrogen as chain transfer agent. Catalyst reactivity with hydrogen to control polymer molecular weight is called its "hydrogen response."

  o In general, the capability of supported chromium catalysts to produce polyethylene of high melt index (low molecular weight) is limited. The hydrogen response of most chromium catalysts is low. Polymer molecular weight may be controlled by chemical modification of the basic chromium catalyst and judicious choice of polymerization temperature and ethylene concentration.

- Control of polydispersity. Though each type of catalyst provides polyethylene with a characteristic range of molecular weight distributions, measures can be taken to expand the range of achievable polydispersities. In general, polyethylene with broad molecular weight distribution provides a good balance of ease of processing while retaining desirable mechanical properties (modulus, toughness, etc.).

- If a copolymer such as VLDPE or LLDPE is the target resin, satisfactory comonomer incorporation must be achieved. This is manifested by the amount of comonomer incorporated (evidenced by density) and the distribution of comonomer in the polymer (evidenced by composition distribution). In general, supported chromium oxide catalysts incorporate comonomer more easily than Ziegler-Natta catalysts.

- Must have proper rate of polymerization, *i.e.*, its "kinetic profile" must fit the process selected. Representative types of kinetic profiles are shown in Figure 1.11.

Additional details will be provided in subsequent chapters on the composition and functioning of transition metal catalysts.

Transition metal catalysts are crucial to the production of polyethylene. Indeed, it would not be practical to produce linear versions of polyethylene without these catalysts. It is difficult to imagine a world without products made from these versatile polymers in our homes, vehicles and workplaces. Ziegler-Natta and supported chromium catalysts will continue to be the dominant catalysts for LLDPE and HDPE for the foreseeable future. However, as single-site catalyst

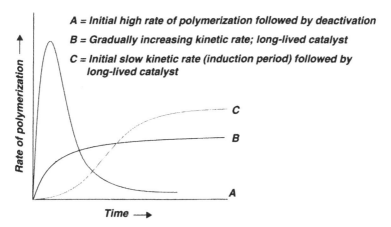

Figure 1.11 Rates of polymerization with transition metal catalysts may be quite different because of accessibility of active centers, presence of poisons, mechanism of activation, etc. Kinetic profile must be accommodated by process conditions (see Chapter 7).

Table 1.3 Principal characteristics of transition metal-containing catalysts for polyethylene production.

	Ziegler-Natta	Metal Oxide Supported	Single-site
Commonly used transiton metals:	Mostly Ti; small amount of V	Mostly Cr; small amount of Mo	Primarily Zr and Ti
Catalyst supports:	$MgCl_2$, SiO_2	SiO_2, SiO_2-Al_2O_3, Al_2O_3, $AlPO_4$	usually not supported
Typical cocatalyst(s):	TEAL	None for 1st generation Phillips	MAO, MMAO and boranes
Primary commercial polymers produced:	LLDPE, VLDPE, HDPE	HDPE, LLDPE	LLDPE, VLDPE
Typical polydispersity range:	4–6	8–20	2–3

technologies mature, they will increase in importance and complement Ziegler-Natta and chromium catalysts in manufacture of polyethylene. A summary of characteristics of the various transition metal catalysts is provided in Table 1.3.

References

1. FM McMillan, *The Chain Straighteners*, MacMillan Publishing Company, London (1979).
2. DM Kiefer, *Today's Chemist at Work*, June 1997, p 51.
3. RB Seymour and T Cheng, *History of Polyolefins*, D. Reidel Publishing Co., Dordrecht, Holland, 1985.
4. RB Seymour, *Advances in Polyolefins*, Plenum Press, New York, 3 (1985).
5. C. Lee and B. Singh, Chemical Marketing Resources, Webster, TX, personal communication, June, 2009.
6. MB Smith and J March, *March's Advanced Organic Chemistry*, John Wiley & Sons, New York, 5th ed., p 74 (2001).
7. B Morris, *International Conference on Polyolefins*, Society of Plastics Engineers, Houston, TX, February 25–28, 2007.
8. RD Jester, *International Conference on Polyolefins*, Society of Plastics Engineers, Houston, TX, February 25–28, 2007.
9. Anon., APEL® *Cyclo Olefin Copolymer* Product Sheet, Mitsui Chemicals Americas, Inc., 2007.
10. MP McDaniel, Handbook of Heterogeneous Catalysis, 1st ed, 1997, G Ertl, H. Knozinger and J. Weitkamp, (editors), VCH Verlagsgesellschaft, Weinheim, Vol 5, 2400.

2

Free Radical Polymerization of Ethylene

2.1 Introduction

In Chapter 1, it was mentioned that highly branched low density polyethylene and copolymers made with polar comonomers are produced only by free radical polymerization at very high pressure and temperature. (All other forms of commercially available polyethylene are produced with transition metal catalysts under much milder conditions; see Chapters 3, 5 and 6.) In this chapter we will review how initiators achieve free radical polymerization of ethylene. Low density polyethylene and copolymers made with polar comonomers are produced in autoclave and tubular processes, to be discussed in Chapter 7.

When linear low density polyethylene from the Unipol® gas-phase process became commercially available in 1975 (1), predictions of the demise of LDPE were widespread.

Though linear low density polyethylene of the period had better mechanical properties and, at that time, could be produced at lower cost, it could not match LDPE's ease of processing and optical properties (especially clarity). Linear low density polyethylene did indeed displace LDPE in many applications. However, LDPE not only survived, it actually grew (2, 3), albeit at a slower pace than other forms of polyethylene. Though more than 75 years have elapsed since its discovery, LDPE from the high pressure process remains a mainstay in the polyolefins industry. After a prolonged period with little or no new capacity, significant projects to construct new plants for production of LDPE have been announced in recent years (4, 5). These newer facilities have much greater capacity (>300,000 tons/y) than earlier plants. It has been reported (3, 6) that economies of scale from modern large-scale plants result in lower costs for LDPE compared to LLDPE, a reversal of the situation when gas phase linear low density polyethylene emerged in the 1970s.

2.2 Free Radical Polymerization of Ethylene

Origins of free radical polymerization of ethylene to produce LDPE were discussed in Chapter 1, stemming from the seminal work of chemists at ICI in the early 1930s. Agents that foster free radical polymerization of ethylene are called "initiators" and sometimes "catalysts." (The latter is not technically correct, since the agents are consumed in the process.) Organic peroxides are the most commonly used initiators for free radical polymerization of ethylene.

Because of the extremely high pressures (15,000 to 45,000 psig), ethylene exists in the liquid phase and polymerization occurs in solution. Owing to high temperatures (typically >200 °C), polyethylene is also dissolved in monomer and the reaction system is homogeneous. LDPE precipitates only after the reaction mass is cooled in post-reactor separation vessels. Relative to other processes, reactor residence times are very short (<30 seconds for the autoclave process and <3 min for the tubular process) (7).

In the polymerization reactor, organic peroxides dissociate homolytically to generate free radicals. Polymerization of ethylene proceeds by a chain reaction. Initiation is achieved by addition of a free radical to ethylene. Propagation proceeds by repeated additions of monomer.

Termination may occur by combination (coupling) of radicals or disproportionation reactions. Chain transfer takes place primarily by abstraction of a proton from monomer or solvent by a macroradical. A low molecular weight hydrocarbon, such as butane, may be used as chain transfer agent to lower molecular weight. Initiation, propagation, termination and chain transfer reactions are shown in Figure 2.1. Termination reactions illustrated in Figure 2.1 show that the end groups in LDPE are most commonly a vinyl group or an ethyl group. Mechanisms for formation (called "backbiting") of n-butyl and 2-ethylhexyl branches are shown in Figures 2.2 and 2.3.

In addition to the use of chain transfer agents, molecular weight may also be varied by adjusting pressure and temperature. Higher pressures lead to higher molecular weight. Branching tends to increase at higher temperatures.

Initiation:

$$ROOR \longrightarrow 2\ RO\cdot$$

(See Figure 2.4 for structures of organic peroxides)

$$2\ RO\cdot + CH_2=CH_2 \longrightarrow ROCH_2CH_2\cdot$$

Propagation:

$$ROCH_2CH_2\cdot + (x+1)CH_2=CH_2 \longrightarrow ROCH_2CH_2(CH_2CH_2)_xCH_2CH_2\cdot \equiv R_p\cdot$$

Termination:

Coupling: $R_p\cdot + R_p\cdot \longrightarrow R_p\text{-}R_p$

Disproportionation: $\sim CH_2CH_2\cdot + \cdot CH_2CH_2 \sim \longrightarrow \sim CH=CH_2 + CH_3CH_2 \sim$

Chain Transfer:

$$\sim CH_2CH_2\cdot + CH_2=CH_2 \longrightarrow \sim CH=CH_2 + CH_3CH_2\cdot$$

$$\sim CH_2CH_2\cdot + CH_2=CH_2 \longrightarrow \sim CH_2CH_3 + CH_2=CH\cdot$$

$$\sim CH_2CH_2\cdot + R'H^* \longrightarrow \sim CH_2CH_3 + R'\cdot$$

* $R'H$ = solvent, CTA, etc.

Figure 2.1 Free radical polymerization of ethylene.

Intramolecular transfer of a radical from a terminal to an internal carbon atom is called "backbiting" and results in short chain branching. This usually occurs on the fifth carbon atom from the macroradical terminus (δ to the radical), as illustrated in Figures 2.2 and 2.3. Ethyl, butyl groups and 2-ethylhexyl groups comprise most of the short chain branching in LDPE (8, 9). Intermolecular transfer between a radical and an internal carbon of another chain results in long chain branching, a characteristic feature of LDPE.

Chain propagation during copolymerization of ethylene with polar comonomers can proceed in several ways depending on the nature of the macroradical end group and the monomer being added, illustrated with vinyl acetate in eq 2.1–2.4:

Self-propagation:

$$\sim CH_2CH_2^{\cdot} + CH_2{=}CH_2 \xrightarrow{\;k_{11}\;} \sim CH_2CH_2CH_2CH_2^{\cdot} \tag{2.1}$$

Cross-propagation:

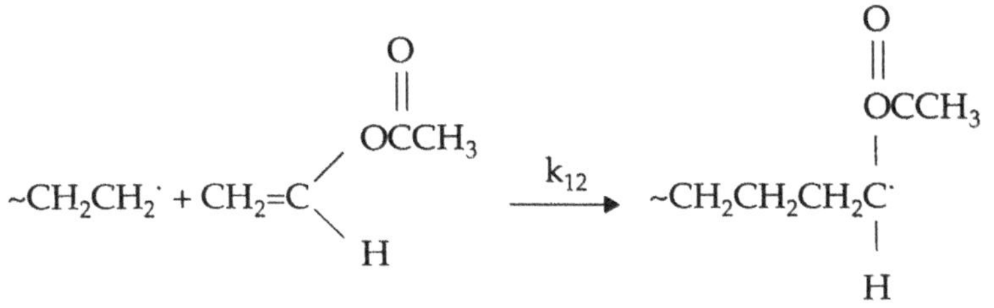

$$\tag{2.2}$$

Cross-propagation:

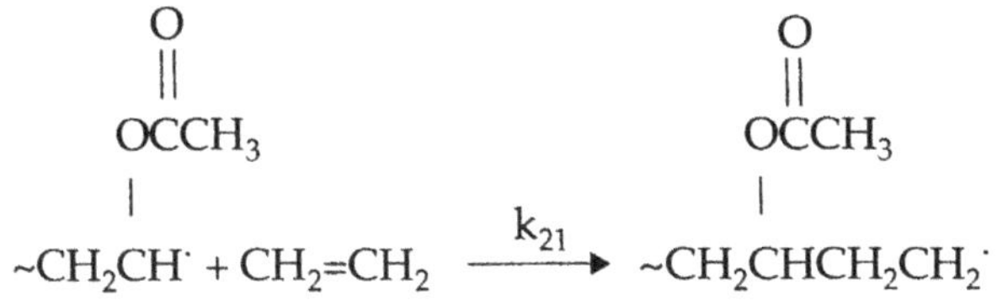

$$\tag{2.3}$$

Self-propagation:

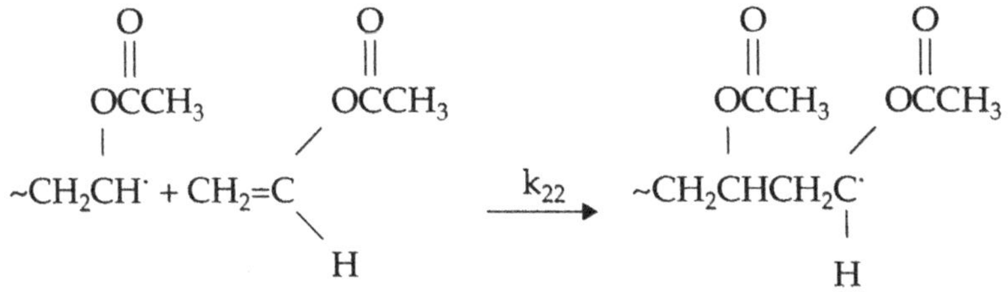

$$\tag{2.4}$$

The ratio of the reaction rate (k_{11}) of an ethylene terminus with ethylene monomer to the reaction rate (k_{12}) of an ethylene terminus with vinyl acetate is defined as the reactivity ratio (r_1):

$$r_1 = k_{11}/k_{12} \tag{2.5}$$

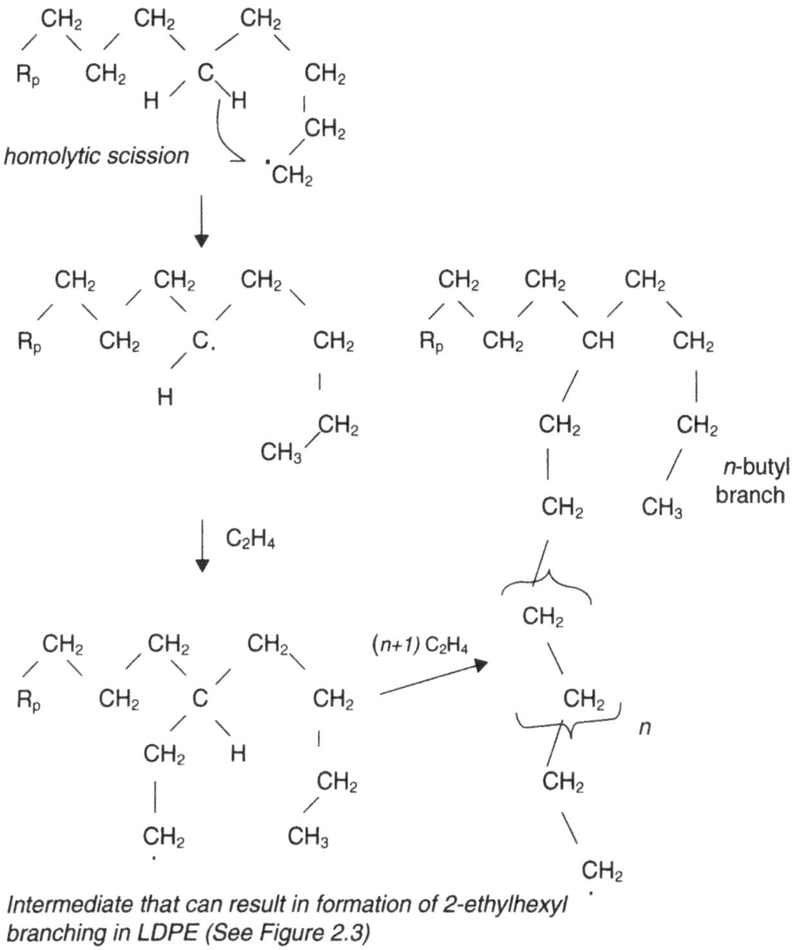

Intermediate that can result in formation of 2-ethylhexyl branching in LDPE (See Figure 2.3)

Figure 2.2 Mechanism of "backbiting" in formation of short chain branching initiated by attack of radical on a δ carbon-hydrogen bond. In the reaction above, homolytic bond scission occurs resulting in a free radical on the 5th carbon atom and an *n*-butyl branch. R_p is a polymeric alkyl group.

Similarly, the ratio of the reaction rate (k_{21}) of a vinyl acetate terminus with ethylene monomer to the reaction rate (k_{22}) of a vinyl acetate terminus with vinyl acetate provides a different reactivity ratio (r_2):

$$r_2 = k_{21}/k_{22} \tag{2.6}$$

Reactivity ratios are indicative of each monomer's tendency to self-propagate or cross-propagate and determine the composition distribution of the polymer.

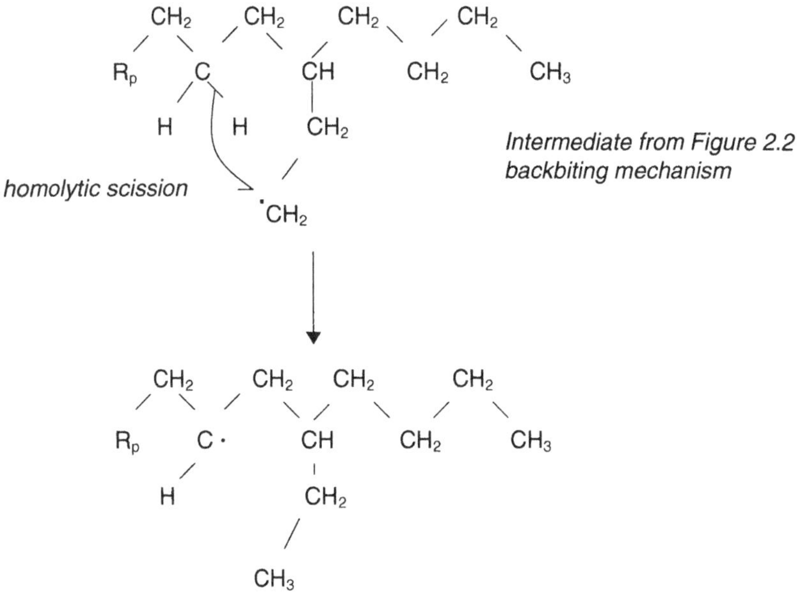

Figure 2.3 Mechanism of formation of 2-ethylhexyl branch in LDPE. As in backbiting mechanism for n-butyl group formation, homolytic scission of a CH bond occurs down the chain. R_p is a polymeric alkyl group.

If $r_1 > 1$, ethylene tends to self-propagate. If $r_1 < 1$, copolymerization is favored. If $r_1 \sim r_2 \sim 1$, the monomers have nearly identical reactivities and comonomer incorporation is highly random. This means that the composition of the copolymer will closely reflect the proportions of ethylene and comonomer charged to the reactor. For EVA, the ethylene reactivity ratio and reactivity ratio for vinyl acetate are very close ($r_1 = 0.97$ and $r_2 = 1.02$), which translates into uniform distribution of VA in the copolymer (10).

Reactivity ratios are important in determining reactor "feed" composition of ethylene and comonomer required to produce a copolymer with the target comonomer content. Because the relative proportion of comonomer changes as polymerization proceeds, adjustment of comonomer feed with time may be necessary. A detailed discussion of the derivation of reactivity ratios for copolymerizations has been provided by Stevens (11).

In addition to potential hazards of handling peroxides (see section 2.3), ethylene itself can decompose violently under the extreme conditions used in high pressure processes for free radical polymerization of ethylene. Anyone living or working within earshot of an LDPE manufacturing facility is familiar with the occasional boom that accompanies the bursting of rupture discs from what is innocuously called a "decomp." This is a result of spontaneous decomposition of ethylene to carbon, hydrogen and methane as depicted in simplified eq 2.7.

$$2CH_2=CH_2 \rightarrow 3C + CH_4 + 2H_2 \tag{2.7}$$

This reaction occurs at ≥ 300 °C and is highly exothermic and releases a pressure pulse that bursts the discs. Even if the standard operating temperature is well below 300 °C, localized "hot spots" from the high heat of polymerization (~24 kcal/mole) can initiate decomposition. Fortunately, decomposition of ethylene is relatively rare. Nevertheless, engineering design must accommodate "decomps." Instrumentation must be capable of detecting excessive pressure or exotherms within microseconds of the event. Mitigating measures (12–14) include:

- rapid depressurization of reactor systems
- injection of nitrogen and water into gases vented upon disc-rupture

2.3   Organic Peroxide Initiators

As discussed above, high pressure ethylene polymerization is usually initiated by organic peroxides. Peroxyesters and dialkyl peroxides are the most common initiators. Organic peroxide initiators are typically clear, colorless liquids and readily undergo homolytic scission to generate free radicals. For safety, some are supplied only in solution. Though often supplied in odorless mineral spirits for LDPE producers, other solvents such as water and mineral oil may also be used. Solvents are also called "phlegmatizers" by suppliers of industrial organic peroxides. Structures of key organic peroxides used in free radical polymerization of ethylene are shown in Figure 2.4.

Organic peroxides are prone to violent decomposition. Safety must be of utmost importance in handling these products. Decomposition can be caused by several factors, but overheating and exposure to metallic impurities (which catalyze decomposition) are the most common. Suppliers of organic peroxides have published "self-accelerating decomposition temperature" (SADT) data. SADT is defined as the lowest temperature at which self-accelerating decomposition may occur. Supplier recommended maximum storage temperatures are typically 15–20 °C below the SADT.

Decomposition of organic peroxides to free radicals follows first order kinetics. Half life ($t_{1/2}$) is defined as the time required for half of the organic peroxide to decompose at a certain temperature. Half-life is an important parameter in

Peroxyesters:

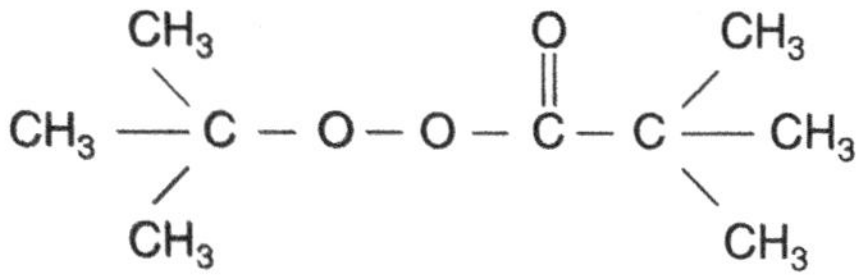

t-butyl peroxypivalate

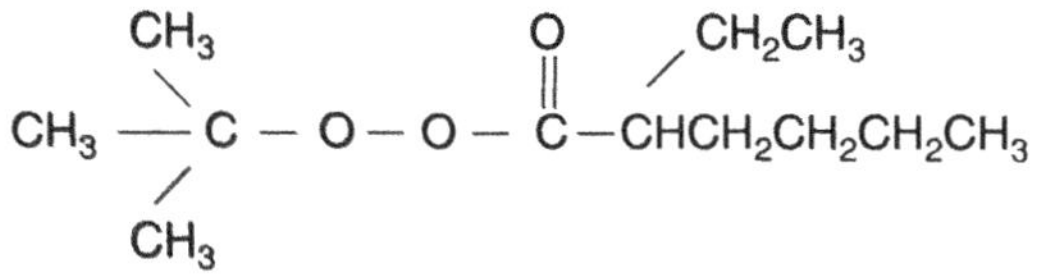

t-butyl peroxy-2-ethylhexanoate

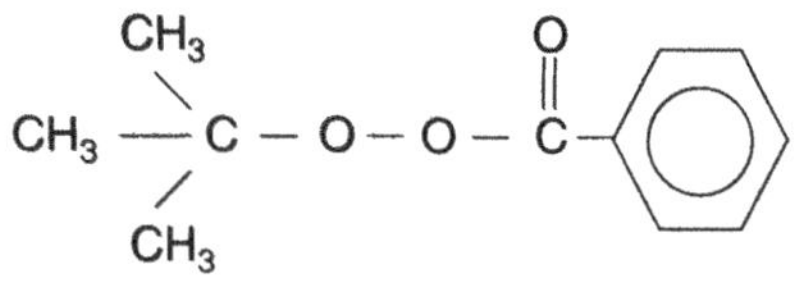

t-butyl peroxybenzoate

Dialkylperoxide:

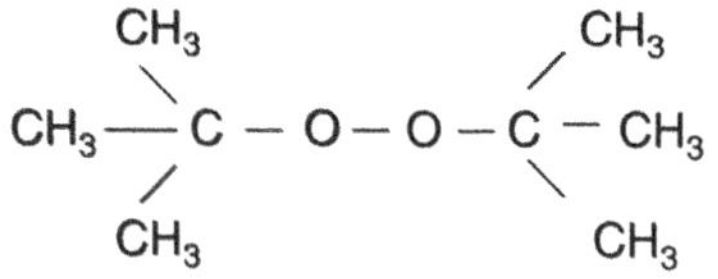

di-t-butyl peroxide

Figure 2.4 Key organic peroxides used to initiate free radical polymerization of ethylene.

selection of the temperature for free radical polymerization of ethylene. As the temperature of $t_{\frac{1}{2}}$ decreases, the activity of the organic peroxide as an initiator increases. The equation for half life for organic peroxides is given in eq (2.8), where k_d is the reaction rate of decomposition of the peroxide. Key organic peroxides used to initiate free radical polymerization of ethylene, their SADTs and temperatures for 0.1 hour $t_{\frac{1}{2}}$ are listed in Table 2.1.

$$t_{\frac{1}{2}} = \ln 2 / k_d = 0.693 / k_d \qquad (2.8)$$

Organic peroxides are supplied globally to manufacturers of LDPE and copolymers produced by free radical polymerization of ethylene. Peroxides are

Table 2.1 Key organic peroxide initiators for LDPE.

Peroxide	Molecular Formula[a]	SADT[b] (°C)	0.1 Hour Half life[b] (t$_{1/2}$) Temperature (°C)
tert-butyl peroxypivalate	$C_9H_{18}O_3$	20[c]	94
tert-butyl peroxy-2-ethylhexanoate	$C_{12}H_{24}O_3$	35	113
tert-butyl peroxybenzoate	$C_{11}H_{14}O_3$	60	142
di-*tert*-butyl peroxide	$C_8H_{18}O_2$	80	164

[a] See Figure 2.4 for structure.

[b] SADT and t$_{1/2}$ data from Akzo Nobel brochure Initiators for Polymer Production, Publication 99-9/1, 1999.

[c] For 75% solution in odorless mineral spirits

transported in refrigerated containers and must be stored at low temperatures as recommended by suppliers. As of this writing, major industrial suppliers are:

- Akzo Nobel
- Arkema Inc. (formerly Atofina)
- Degussa Initiators
- GEO Specialty Chemicals
- LyondellBasell Chemical Company
- NORAC, Inc.

References

1. FJ Karol, *History of Polyolefins* (RB Seymour and T Cheng, editors), D. Reidel Publishing Co., Dordrecht, Holland, p 199, 1985.
2. CE Schuster, *Handbook of Petrochemicals Production Processes*, (RA Meyers, ed.), McGraw-Hill, NY, p 14.55, 2005.

3. MJ Kaus, 2005 *Petrochemical Seminar*, Mexico City (moved from Cancun), November 4, 2005.
4. Anon., *Plastics Engineering*, Society of Plastics Engineers, Brookfield CT, p 39, August, 2006.
5. Anon., *Modern Plastics*, Canon Communications LLC, Los Angeles, CA, p 8, November, 2004.
6. AH Tullo, *Chemical & Engineering News*, p 26, June 2, 2003.
7. CE Schuster, *Handbook of Petrochemicals Production Processes*, (RA Meyers, ed.), McGraw-Hill, NY, p 14.53, 2005.
8. A. Peacock, *Handbook of Polyethylene*, Marcel Dekker, New York, p 46, 2000.
9. KS Whitely, *Ullman's Encyclopedia of Industrial Chemistry*, Wiley-VCH Verlag Gmbh & Co., 2002.
10. MP Stevens, *Polymer Chemistry*, 3rd ed., Oxford University Press, New York, p 196, 1999.
11. *ibid*. p 194, 1999.
12. A-A Finette and G ten Berge, *Handbook of Petrochemicals Production Processes*, (RA Meyers, ed.), McGraw-Hill, NY, p 14.108, 2005.
13. M. Mirra, *Handbook of Petrochemicals Production Processes*, (RA Meyers, ed.), McGraw-Hill, NY, p 14.67, 2005.
14. CE Schuster, *Handbook of Petrochemicals Production Processes*, (RA Meyers, ed.), McGraw-Hill, NY, p 14.52, 2005.

3

Ziegler-Natta Catalysts

3.1 A Brief History of Ziegler-Natta Catalysts

The Ziegler-Natta catalyst is so named in recognition of the pioneering work in the 1950s of Karl Ziegler in Germany and Giulio Natta in Italy. Ziegler discovered the basic catalyst systems that polymerize ethylene to linear high polymers (more below). However, Ziegler's preliminary experiments with propylene were not successful. Ziegler decided to concentrate on expanding knowledge of catalysts in ethylene polymerization and postponed further work on propylene. Natta was then a professor at the Institute of Industrial Chemistry at Milan Polytechnic and a consultant with the Italian company Montecatini. Natta had arranged for a cooperative research and licensing agreement between Ziegler and Montecatini. Through this arrangement, he learned of Ziegler's success with ethylene polymerization and pursued propylene polymerization aggressively. Natta succeeded in producing crystalline polypropylene and determining its crystal structure in early 1954. Ziegler and Natta were jointly awarded the Nobel Prize in Chemistry in 1963 for their work in polyolefins.

Though Ziegler had an enduring interest in metal alkyl chemistry going back to the 1920s, it was of not until the late 1940s that he discovered the "aufbau" (growth) reaction. The aufbau reaction occurs in the essential absence of transition metal compounds and is not capable of producing high molecular weight polyethylene. However, it served as the crucial precursor discovery that culminated with the development of commercially viable Ziegler-Natta catalysts for polyolefins. In the aufbau reaction, triethylaluminum (TEAL) reacts with ethylene *via* multiple insertions to produce long chain aluminum alkyls with an even number of carbon atoms. Under proper conditions, β-elimination occurs resulting in formation of α-olefins. The aluminum hydride moiety resulting from elimination is then able to add ethylene to start a new chain. If the long chain aluminum alkyls are first air-oxidized and then hydrolyzed, α-alcohols are formed. These reactions collectively are called "Ziegler chemistry" (1) and are illustrated in Figure 3.1. Ziegler chemistry forms the basis for present-day production of many millions of pounds of α-olefins, used as comonomers for LLDPE and VLDPE, and α-alcohols, intermediates for detergents and plasticizers.

Ziegler and coworkers at the Max Planck Institut für Kohlenforschung (Coal Research) in what was then Mulheim, West Germany were working to expand the scope and utility of the aufbau reaction. It was during this endeavor in 1953 that they accidentally discovered the "nickel effect." This term stemmed from the observation that nickel in combination with triethylaluminum catalyzes dimerization of ethylene to produce 1-butene. Accounts vary on the source of nickel in the formative experiments. It was ultimately attributed to trace nickel extracted from the surface of the stainless steel reactor in which early reactions were conducted.

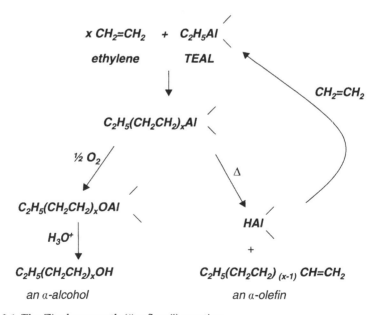

Figure 3.1 The Ziegler growth ("aufbau") reaction.

Ziegler then launched a systematic study of the effects of transition metal compounds combined with aluminum alkyls on ethylene. Heinz Breil, a young graduate student, was assigned responsibility. Though early experiments were disappointing, Breil persisted and eventually combined zirconium acetylacetonate with triethylaluminum and produced linear polyethylene as a white powder. Later, Heinz Martin combined titanium tetrachloride with triethylaluminum and obtained a highly active catalyst for ethylene polymerization. (Indeed, in the initial experiment, the catalyst was so "hot" that the polymer was charred.) Though combinations of titanium compounds and aluminum alkyls have been modified and refined over the past 50+ years, it is remarkable that such combinations capture the essential character of most early 21st century Ziegler-Natta polyethylene catalysts.

The origins of Ziegler-Natta polyolefin catalysts have been authoritatively described by McMillan (2), Seymour (3, 4), Boor (5) and Vandenberg and Repka (6).

3.2 Definitions and Nomenclature

In broad terms, Ziegler-Natta catalysts are defined as combinations of a transition metal compound from Groups 3–12 of the Periodic Table with an organometallic compound from Groups 1, 2 or 13. Each component alone is incapable of converting olefins to high polymers. (Interactions between catalyst and cocatalyst will be addressed in section 3.7 on the mechanism of Ziegler-Natta polymerization.) These dual component catalyst systems are not merely "complexes." Rather, substantial reactions occur between the organometallic and transition metal compounds. Of course, all combinations don't work as polymerization catalysts, but patent claims were made very broad to maximize coverage. Most commercial Ziegler-Natta catalysts are heterogeneous solids, but some, primarily those derived from vanadium compounds, are homogeneous (soluble). After polymerization, the catalyst is interspersed throughout the polymer and cannot be isolated. (Hence, recycle of Ziegler-Natta catalysts is not practical.) In reduced form, the transition metal component is called the "catalyst," while the organometallic part is called the "cocatalyst," or less commonly, the "activator." In most cases, a titanium compound (often titanium tetrachloride, $TiCl_4$) and an aluminum alkyl are used. (Aluminum alkyls and their roles in polymerization will be discussed in Chapter 4.)

Catalyst activity, also called yield, productivity or mileage, may be expressed in several ways. Frequently used units are weight (g, kg or lb) of polyethylene per wt of catalyst, and weight of polyethylene per weight of transition metal per atmosphere of ethylene per hour (written, for example, as g PE/g Ti-atm C_2H_4-h for a titanium catalyst). The latter method is often used in journal literature. Weight per weight of catalyst is typically used in manufacturing operations and is independent of catalyst residence time. (Average catalyst residence times in commercial reactors can range from fractions of a second to several hours, depending on process and catalyst. See Chapter 7.)

A few additional comments are warranted on nomenclature. Since discovery in the 1950s, Ziegler-Natta catalysts have been known by a variety names such as "coordination catalysts," "coordinated anionic catalysts" (originally proposed by Natta), "Ziegler catalysts" (applied mostly to polyethylene) and "Natta catalysts" (applied mostly to polypropylene). The term "Ziegler chemistry" has also been applied to reactions of aluminum alkyls with ethylene (the "aufbau" reaction (1)) in the essential absence of transition metal compounds. In this text, the notation "Ziegler-Natta catalysts" is meant to encompass ethylene polymerizations (and copolymerizations) employing catalysts produced by combining transition metal salts with metal alkyls, generally in keeping with the rationale used by Boor long ago (see p 34 of reference 5). However, single site catalysts are considered here to be separate from Ziegler-Natta catalysts. Reasons for this are discussed below.

Single site ("metallocene") catalysts are considered by some catalyst chemists to be a subset of Ziegler-Natta catalysts, because they involve combinations of transition metal compounds with Group 13 organometallics. As previously noted, relatively few Ziegler-Natta catalysts are homogeneous. In contrast, single site catalysts are homogeneous, produce polyethylene with properties that are quite distinct from polyethylene made with Ziegler-Natta catalysts, and polymerize olefins *via* a different mechanism. For these reasons, single site catalysts will be discussed separately (see Chapter 6).

3.3   Characteristics of Ziegler-Natta Catalysts

Ziegler-Natta catalysts are not pure compounds. Most are heterogeneous inorganic solids, essentially insoluble in hydrocarbons and other common organic solvents, making them difficult to study. In reduced form, Ziegler-Natta catalysts are typically highly colored (from violet to gray to brown) powdery or granular solids. Many smoke or ignite upon exposure to air and may be violently reactive with water. Because Ziegler-Natta catalysts may be rendered inactive ("poisoned") even by traces of oxygen and water, they must be handled under an inert atmosphere (usually nitrogen). Consequently, polyolefin manufacturers typically combine components on site to produce the active Ziegler-Natta catalyst system. In some cases, the polyethylene manufacturer produces its own catalyst and purchases cocatalyst from metal alkyl suppliers (see section 4.2.1). Ziegler-Natta catalyst systems are not storage stable once the two components are combined. Catalyst and cocatalyst are usually combined in the polymerization reactor. Catalyst recipes are highly proprietary and it is often difficult to discern actual practices from what McMillan called the "fairyland of patent applications" (7).

An important characteristic of heterogeneous polyethylene catalysts is the phenomenon of particle replication. Particle size distribution (psd) and morphology of the catalyst are reproduced in the polymer. If the catalyst is finely divided, the polymer will also be fine and may cause handling problems. If the catalyst contains agglomerates of oversized particles, so too will the polymer. Morphology replication is illustrated in Figure 3.2. Figure 3.3 shows how psd of the catalyst is mirrored in the polymer.

Ziegler-Natta catalysts polymerize ethylene under very mild conditions relative to those needed for free radical polymerization. For example, conditions for free radical processes are often >200 °C and pressures of >140 MPa (~20,000 psig). In contrast, Ziegler's group showed that Ziegler-Natta catalysts are capable of polymerizing ethylene at atmospheric pressure and ambient temperature

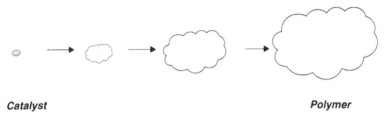

Catalyst   **Polymer**

Figure 3.2 As the polymer particle grows, it assumes the morphology of the catalyst particle. This is referred to as "replication," *i.e.*, a spherical catalyst results in a spherical polymer particle.

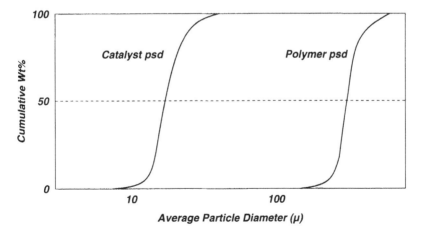

Figure 3.3 Particle size distribution of polymer mirrors that of catalyst particles. In this example, a catalyst with ~40 µ average diameter results in a polymer particle of ~500 µ.

(see p 67 of ref 2). Another profound difference is the microstructure of the resultant polyethylenes. Ziegler-Natta catalysts produce linear polyethylene while the high pressure process yields highly branched polyethylene, as previously shown in Figure 1.2.

3.4   Early Commercial Ziegler-Natta Catalysts

Titanium tetrachloride was the logical choice as the raw material for early Ziegler-Natta catalysts. $TiCl_4$ is a clear, colorless, hygroscopic liquid that fumes upon exposure to ambient air. $TiCl_4$ (aka "tickle 4") was (and still is) manufactured in enormous volumes as a precursor to titanium dioxide used as a pigment for paint. Consequently, $TiCl_4$ was readily available and relatively inexpensive. Also, $TiCl_4$ had been shown by Ziegler and coworkers to produce some of the most active polyethylene catalysts. Though sometimes called a catalyst, it is more accurate to call $TiCl_4$ a "precatalyst," since it must be reduced and combined with a cocatalyst to become active.

Early (1960–1965) commercial Ziegler-Natta catalysts were produced by reduction of $TiCl_4$ using metallic aluminum, hydrogen or an aluminum alkyl. Simplified overall equations are listed below:

$$TiCl_4 + \tfrac{1}{3}\,Al \rightarrow TiCl_3 \bullet \tfrac{1}{3}\,AlCl_3 \text{ [also written } (TiCl_3)_3 \bullet AlCl_3] \quad (3.1)$$

$$2\,TiCl_4 + H_2 \rightarrow 2\,TiCl_3 + 2\,HCl \quad (3.2)$$

$$2\,TiCl_4 + 2\,(C_2H_5)_3Al_2Cl_3 \rightarrow 2\,TiCl_3 \downarrow + 4\,C_2H_5AlCl_2 + C_2H_4 + C_2H_6 \quad (3.3)$$

The predominant product in each case was titanium trichloride (aka "tickle 3"), an active catalyst for olefin polymerization. The preferred cocatalyst was diethylaluminum chloride (DEAC). $TiCl_3$ from eq 3.1 contains co-crystallized aluminum trichloride. $TiCl_3$ from eq 3.3 may contain small amounts of complexed aluminum alkyl. Products from eq 3.1 and 3.2 were supplied commercially by companies such as Stauffer Chemical and Dart (both now defunct). Catalyst from eq 3.3 was manufactured on site by polyolefin producers, usually in an inert hydrocarbon such as hexane.

Activity was poor (500–1000 g polyethylene/g catalyst) with these antiquated catalysts. Early polyethylene producers were required to post-treat polymer to remove acidic chloride- and transition metal-containing residues that could cause discoloration of polymer and corrosion of downstream processing equipment. In the mid-1960s, ball-milling of catalysts was introduced and resulted in increased surface area and higher activity. Catalysts from eq 3.1–3.3 are now largely obsolete. However, $TiCl_3 \bullet \tfrac{1}{3}\,AlCl_3$ (from eq 3.1) is still supplied commercially, but at greatly reduced volumes.

$TiCl_3$ is a highly colored solid that exists in several crystalline forms, designated as alpha (α), beta (β), gamma (γ) and delta (δ). The α, γ and δ forms have layered crystal structures and are violet. The β form has a linear structure and is brown.

3.5 Supported Ziegler-Natta Catalysts

In non-supported catalysts, most active centers (>95%) become encased within the growing polymer particle and thereby become unavailable for additional polymer formation. This results in low catalyst activity. A major improvement occurred in the early 1970s when supported Ziegler-Natta catalysts began to emerge. Leading polyethylene producers of the time (Shell, Solvay & Cie, Hoechst, Mitsui and Montecatini Edison) developed many of these catalysts (8). Of course, most have morphed into present-day companies, such as LyondellBasell and INEOS.

Supported catalysts result in dispersed active centers that are highly accessible. Catalyst activity is greatly increased (>5,000 g polyethylene/g catalyst). TEAL is the preferred cocatalyst for supported Ziegler-Natta catalysts. Transition metal residues in polyethylene produced with modern supported catalysts are very low (typically <5 ppm), obviating post-reactor treatment of polymer.

Supports also make it possible to control psd and morphology of the catalyst, which is reflected in the polymer, previously illustrated in Figures 3.2 and 3.3. For example, a spherical catalyst particle of 40–60 microns (μ) average diameter in the Unipol® gas phase process will grow into a spherical polymer particle of 500–1000 μ average diameter, with a residence time of about 4 hours (9). Proper control of catalyst psd is desirable because it translates into narrower polymer psd, thereby minimizing large (and small) particles. Because on-site industrial polymer transfer is achieved primarily by pneumatic conveyance, large particles can lead to flow problems, such as plugged transfer lines. Fine particulates may cause clogged filters and heighten the possibility of dust explosions. Morphology control increases bulk density of the finished resin and improves fluidization dynamics in gas phase processes.

Many inorganic compounds were tested as supports, but magnesium salts and silica provided the most serviceable catalysts. Magnesium compounds such as MgO, $Mg(OH)_2$, HOMgCl, ClMgOR and $Mg(OR)_2$, have been used, but anhydrous $MgCl_2$ is the most widely employed in commercial Ziegler-Natta catalysts. Active centers are chemisorbed on the surface of the magnesium compound. Chien described $MgCl_2$ as the "ideal support" for titanium active centers and summarized several reasons for the desirability of $MgCl_2$ as a support (10). For example, Chien cited the similarity of crystal structures between $MgCl_2$ and $TiCl_3$. Worldwide, $MgCl_2$-supported Ziegler-Natta catalysts have become the most important catalysts for olefin polymerization.

Silica is sometimes called a "carrier" since the catalyst may be simply deposited onto the support. (Not so for supported chromium catalysts, where the catalyst is strongly bonded to the support; see Chapter 5.) Even in cases of simple deposition of catalyst, however, the psd and morphology of the silica dictates the psd and shape of the polymer particle.

3.6 Prepolymerized Ziegler-Natta Catalysts

A technique called "prepolymerization" is practiced with selected psd controlled catalysts used in slurry and gas phase processes. The catalyst is suspended in a suitable solvent (usually a C_3-C_7 alkane) and exposed to cocatalyst, ethylene, and, optionally, comonomer and hydrogen under very mild conditions in a separate, smaller reactor (11). Prepolymerization is allowed to proceed until the original catalyst comprises 5–30% of the total weight of the composition. Prepolymerized catalysts have limited storage stability and are ordinarily introduced without delay to the large-scale reactor. Prepolymerization provides several advantages:

- Preserves catalyst morphology by making the catalyst particle more "robust," *i.e.*, less prone to fragmentation (which may produce undesirable fines)
- Insulates the catalyst particle, thereby reducing sensitivity to heat during initial stages of polymerization
- Increases resin bulk density

3.7 Mechanism of Ziegler-Natta Polymerization

Despite passage of more than 57 years since the basic discoveries, the mechanism of Ziegler-Natta polymerization is still not fully understood. As in all chain-growth polymerizations (12), the basic steps are initiation, propagation and termination (chain transfer).

Cossee and Arlman (13, 14) were among the first to propose a comprehensive mechanism for Ziegler-Natta catalysis and supported their proposals with molecular orbital calculations. The Cossee-Arlman proposal involves a "migratory alkyl transfer" (15) and, with some refinements, remains the most widely cited mechanism for Ziegler-Natta catalysis. A summary is presented below. (For more details, see references 5, 12, 16 and 17.)

The reduced form of titanium is octahedral and contains open coordination sites (□) and chloride ligands on crystallite edges. Initiation begins by formation of an active center, believed to be a titanium alkyl. Alkylation by TEAL cocatalyst produces an active center:

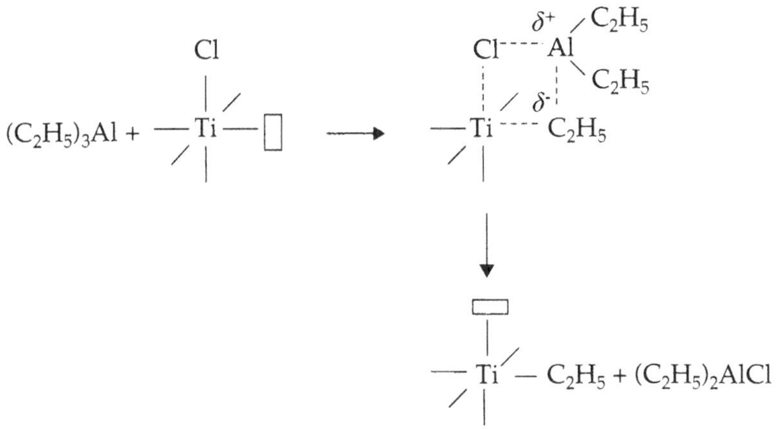

$$(C_2H_5)_3Al + \quad -Ti- \square \quad \longrightarrow \quad -Ti \cdots C_2H_5 \tag{3.4}$$

$$-Ti - C_2H_5 + (C_2H_5)_2AlCl$$

The alkyl migrates (rearranges) such that an open coordination site moves to a crystallite edge position. Coordination of an ethylene monomer occurs to create a π-complex as in eq 3.5. Subsequent addition across ethylene results in the propagating species:

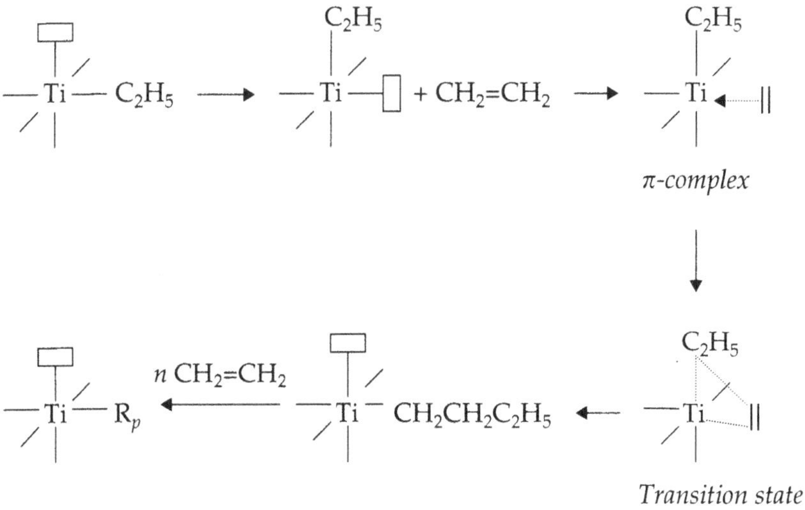

$$-Ti - C_2H_5 \longrightarrow -Ti-\square + CH_2=CH_2 \longrightarrow -Ti \cdots \|$$

π-complex

$$\tag{3.5}$$

$$-Ti-R_p \xleftarrow{\;n\;CH_2=CH_2\;} -Ti \; CH_2CH_2C_2H_5 \longleftarrow -Ti \cdots \|$$

Transition state

$R_p = (-CH_2CH_2)_{n+1}\,C_2H_5$, *a polymer chain*

Because titanium-carbon σ-bonds are known to be unstable, a different mechanism that invokes coordination of the aluminum alkyl to the titanium alkyl has been postulated. It is suggested that the titanium alkyl is stabilized by association with the aluminum alkyl. Coordination also accommodates the well-known propensity of aluminum alkyls to associate (18). This is known as the "bimetallic mechanism," and essential features were originally proposed by Natta and other workers in the early 1960s (19). Basic steps are similar to the Cossee-Arlman mechanism. The principal difference is participation of the aluminum alkyl. However, polymerization is still believed to occur by insertion of C_2H_4 into the Ti-C bond (rather than the Al-C bond). Key steps are illustrated in eq 3.6 below:

$$(3.6)$$

Termination occurs primarily through chain transfer to hydrogen, that is, hydrogenolysis of the R_p-Ti bond as in eq 3.7. The titanium hydride may add ethylene to produce another active center for polymerization.

$$(3.7)$$

Chain termination may also occur by β-elimination with hydride transfer to titanium (eq 3.8), by β-elimination with hydride transfer to monomer (eq 3.9) and chain transfer to aluminum alkyl (eq 3.10).

$$-\overset{|}{\underset{|}{Ti}}-H + CH_2=CHR_p \qquad (3.8)$$

$$-\overset{|}{\underset{|}{Ti}}-CH_2CH_3 + CH_2=CHR_p \qquad (3.9)$$

$$(C_2H_5)_3Al + -\overset{|}{\underset{|}{Ti}}-R_p \rightarrow -\overset{|}{\underset{|}{Ti}}----R_p \rightarrow -\overset{|}{\underset{|}{Ti}}-\square + (C_2H_5)_2AlR_p \qquad (3.10)$$

The aluminum alkyl product from eq 3.10 containing the polymeric chain (R_p) will undergo hydrolysis or oxidation/hydrolysis when the resin is exposed to ambient air, similar to the chemistry depicted in Figure 3.2. This chemistry results in polymer molecules with methyl and ~CH_2OH end groups, respectively. However, concentrations are miniscule, since the vast majority of chain termination occurs by eq 3.7–3.9.

In chain transfer/terminations illustrated in eq 3.7–3.10, the component containing the transition metal is still an active catalyst. Thus, each active center may produce hundreds or thousands of polymer chains.

The mechanism for polymerization of propylene using Ziegler-Natta catalysts is analogous to that discussed in section 3.7 with ethylene. However, unlike ethylene, propylene can be said to have "head" and "tail" portions and regiochemistry can vary. More importantly, the orientation (stereochemistry) of the methyl group in the polymer has a dramatic effect on polymer properties. These factors make polymerization of propylene (and other α-olefins) more complex (17).

Ziegler-Natta catalysts are the most important transition metal catalysts for production of polyethylene as well as other poly-α-olefins. Indeed, at this writing, it would not be practicable to manufacture the quantities of stereoregular polypropylene needed for the global market without Ziegler-Natta catalysts. This may change as single site catalyst technology continues to evolve, but Ziegler-Natta catalysts will remain essential to polyolefin manufacture well into the 21st century.

References

1. JR Zietz, Jr., GC Robinson and KL Lindsay, *Comprehensive Organometallic Chemistry*, Vol 7, p 368, 1982.
2. FM McMillan, *The Chain Straighteners*, MacMillan Press, London, 1979.
3. RB Seymour and T Cheng, *History of Polyolefins*, D. Reidel Publishing Co., Dordrecht, Holland, 1986.
4. RB Seymour and T Cheng (editors), *Advances in Polyolefins*, Plenum Press, New York, 1987.
5. J Boor, Jr., *Ziegler-Natta Catalysts and Polymerizations*, Academic Press, Inc., 1979.
6. EJ Vandenberg and BC Repka, *High Polymers*, (ed. CE Schildknecht and I Skeist), John Wiley & Sons, *29*, p 337, 1977.
7. Reference 3, p xi.
8. FJ Karol, *Encyclopedia of Polymer Science and Technology*, Supp Vol 1, p 120 (1976).
9. FJ Karol, *Macromol. Symp.*, 1995, *89*, 563.
10. JCW Chien, *Advances in Polyolefins*, Plenum Press, New York, p 256, 1987.
11. T Korvenoja, H Andtsjo, K Nyfors and G Berggren, *Handbook of Petrochemicals Production Processes*, McGraw-Hill, p 14.18, 2005.
12. MP Stevens, *Polymer Chemistry*, 3rd ed., Oxford University Press, New York, p 11, 1999.
13. P Cossee, *J. Catal.* 1964, *3*, 80.
14. EJ Arlman and P Cossee, *J Catal.* 1964, *3*, 99.
15. FA Cotton, G Wilkinson, CA Murillo and M Bochmann, *Advanced Inorganic Chemistry*, 6th ed., John Wiley & Sons, New York, p 1270, 1999.
16. JP Collman, LS Hegebus, JR Norton and RG Finke, *Principles and Applications of Organo-transition Metal Chemistry*, University Science Books, Sausalito, CA, p 100, 1987.
17. BA Krentsel, YV Kissin, VJ Kleiner and LL Stotskaya, *Polymers and Copolymers of Higher α-Olefins*, Hanser/Gardner Publications, Inc., Cincinnati, OH, p 6, 1997; YV Kissin, *Alkene Polymerizations with Transition Metal Catalysts*, Elsevier, The Netherlands, 2008; G. Cecchin, G. Morini and F. Piemontesi, *Kirk-Othmer Encyclopedia of Chemical Technology*, Wiley Interscience, New York, Vol 26, p 502, 2007.
18. T Mole and EA Jeffery, *Organoaluminium Compounds*, Elsevier Publishing Co., Amsterdam, p 95, 1972.
19. J Boor, Jr., *Ziegler-Natta Catalysts and Polymerizations*, Academic Press, Inc., p 334, 1979.

4

Metal Alkyls in Polyethylene Catalyst Systems

4.1 Introduction

Metal alkyls, defined as organometallic compounds containing at least one carbon-to-metal σ-bond, are essential to the performance of Ziegler-Natta catalysts and, as demonstrated in Chapter 3, are intimately involved in the mechanism for Ziegler-Natta polymerization. Most of the metal alkyls commonly used with industrial polyethylene catalysts ignite spontaneously when exposed to air (*i.e.*, they are pyrophoric) and are explosively reactive with water. Metal alkyls are also used with selected chromium catalysts and with virtually all single site catalysts (see Chapters 5 and 6). For the polyethylene industry, the most important types are aluminum alkyls and magnesium alkyls. Modern ZN catalysts employ aluminum alkyls chiefly as cocatalysts, while magnesium alkyls are used almost exclusively as raw materials for the production of supported catalysts. This chapter will survey the principal metal alkyls that are employed in industrial polyethylene processes and briefly review aspects of safety and handling of these highly reactive chemicals.

Though aluminum and magnesium alkyls are most important, lithium, boron and zinc alkyls are also useful in niche polyethylene applications. While in-depth discussions of production, properties and applications of metal alkyls are outside the scope of this book, key industrial methods for production of metal alkyls will be briefly discussed. However, extensive reviews are available where more information may be obtained (1–8). Metallocenes are π-bonded organometallics and will be discussed in the context of single-site catalysts in Chapter 6.

4.2 Aluminum Alkyls in Ziegler-Natta Catalysts

Aluminum alkyls have been produced commercially since 1959 using technology originally licensed by Karl Ziegler. Aluminum alkyls are typically pyrophoric and explosively reactive with water (3, 9, 10). Considering such hazards, it is remarkable that thousands of tons of aluminum alkyls are produced each year and have been supplied for decades to the polyolefins industry worldwide with relatively few safety incidents. (However, see section 4.7.)

The first large-scale production of an aluminum alkyl via Ziegler chemistry was by Texas Alkyls, Inc. (then a joint venture of Hercules and Stauffer Chemicals) in November of 1959. Karl Ziegler's revolutionary "direct process" was developed in the early 1950s not long after the other exciting discoveries in olefin polymerization that were made in his laboratories. Ziegler's direct process effectively involved reaction of aluminum metal, olefin and hydrogen to produce trialkylaluminum compounds. (This is necessarily an oversimplification of the direct process. Please see references 1, 3, 10 and 11 for more details). Key reactions involved in the Ziegler direct process of triethylaluminum are shown in eq 4.1 and 4.2 below:

Hydrogenation: $\qquad 2\ (C_2H_5)_3Al + Al + 3/2\ H_2 \rightarrow 3\ (C_2H_5)_2AlH \qquad$ (4.1)

Addition: $\qquad 3\ C_2H_4 + 3\ (C_2H_5)_2AlH \rightarrow 3\ (C_2H_5)_3Al \qquad$ (4.2)

Adding equations 4.1 and 4.2 gives the overall reaction for the direct process shown in eq 4.3:

Overall Reaction: $\qquad 3\ C_2H_4 + Al + 3/2\ H_2 \rightarrow (C_2H_5)_3Al \qquad$ (4.3)

However, the reaction shown in eq 4.3 does not take place in the absence of "pre-formed" triethylaluminum.

Triethylaluminum has also been produced industrially by the so-called "exchange process" illustrated in eq 4.4 with triisobutylaluminum and ethylene:

$$(isoC_4H_9)_3Al + 3\ C_2H_4 \rightarrow (C_2H_5)_3Al + 3\ isoC_4H_8 \uparrow \qquad (4.4)$$

Both the direct and exchange processes may be run continuously. However, economics favor the direct process and the direct product also has fewer contaminants. The exchange process is no longer used for triethylaluminum, but is still used for specialty products such as "isoprenylaluminum" (from reaction of triisobutylaluminum and isoprene, 3).

The Ziegler direct process technology far surpassed historical methods for synthesis of trialkylaluminum compounds. Excellent conversions and yields are obtained with relatively little waste, since all raw materials are incorporated into the product (3). Texas Alkyls was acquired by Akzo Chemicals (now Akzo Nobel) in 1992.

Major suppliers for aluminum alkyls as of 2010 are:

- Akzo Nobel (formerly Texas Alkyls, Inc.)
- Albemarle (formerly Ethyl Corp.)
- Chemtura (formerly Crompton, Witco and Schering)

These companies supply aluminum alkyls globally. Akzo Nobel and Albemarle have their principal aluminum alkyl manufacturing facilities in the USA. Chemtura's main site is in Germany. A few regional suppliers, such as Tosoh Finechem Corporation in Japan, also manufacture aluminum alkyls but have lower capacities and narrower product range.

Major suppliers of metal alkyls have joint ventures and satellite plants around the world. Some of the joint ventures and satellite plants have manufacturing capabilities (but with only a few major products). Others have only repackaging and solvent blending facilities to serve regional customers using products imported in bulk from the principal manufacturing plant.

A joint venture between Albemarle and SABIC was recently announced. The joint venture, to be called Saudi Organometallic Chemicals, will have a capacity of about 6000 tons per year of triethylaluminum (12). Capacity for other products was not disclosed.

More than 20 aluminum alkyls are presently offered in the merchant market. As of this writing, most of the high-volume products are priced between about $5 and $10 per pound. Exceptions include trimethylaluminum (which is produced by a costly multi-step process (13)) and diethylaluminum iodide (which requires expensive iodine). Triethylaluminum (TEAL) is the most important aluminum alkyl and is sold globally in multi-million pound per year quantities. Large amounts of triethylaluminum are used in production of polypropylene. Chlorinated aluminum alkyls, such as diethylaluminum chloride (DEAC) and ethylaluminum sesquichloride (EASC), are less costly than triethylaluminum. However, DEAC and EASC do not perform well with some modern supported catalysts (especially polypropylene catalysts) and have declined in importance since the 1980s.

Triisobutylaluminum (TIBAL) is a commercially available trialkylaluminum that performs comparably to triethylaluminum with many Ziegler-Natta catalysts and typically costs less per pound than triethylaluminum. So, why isn't TIBAL the number one selling aluminum alkyl? The reason is that, if other factors are equal, polyolefin manufacturers buy on the basis of *contained* aluminum. Since triethylaluminum contains about 70% more aluminum on a molar basis, TIBAL actually costs substantially *more* than triethylaluminum based on aluminum content, accounting for the dominance of triethylaluminum. Table 4.1 illustrates the differences in cost of *contained* aluminum when prices (per lb) of triethylaluminum and triisobutylaluminum are assumed to be identical.

As illustrated above, selection of cocatalyst is often predicated on cost. In some cases, however, use of an alternative cocatalyst may transcend the cost factor. This could be because the alternative cocatalyst provides enhanced polymer properties or improved process performance. For example, use of TMAL as cocatalyst in place of TEAL in a gas phase process has been shown to provide linear low density polyethylene with lower hexane extractables and superior film tear strength (14). Ultrahigh molecular weight polyethylene and polyethylene with broader molecular weight distribution can be produced using "isoprenylaluminum" as cocatalyst (15–17).

Aluminum alkyls fulfill several roles in the Ziegler-Natta catalyst system as described in the following sections.

Table 4.1 Comparative cost of selected trialkylaluminum compounds.

Product	Assumed Price* ($/lb)	Typical Al Content (wt %)	Cost of Contained Al ($/lb)	Cost of Contained Al (% relative to TEAL)
trimethylaluminum**	100	36.8	271.74	628
triethylaluminum	10	23.1	43.29	100
triisobutylaluminum	10	13.6	73.53	170
tri-*n*-hexylaluminum	10	9.8	102.04	236

* For illustration only; not actual commercial prices. Contact major manufacturers to obtain current bulk pricing.

** TMAL is manufactured by a different process than other R_3Al and is much more expensive. See reference 13.

4.2.1   Reducing Agent for the Transition Metal

This function can be effectively illustrated with a catalyst synthesis used in an early commercial polypropylene process, now obsolete. The catalyst system employed ethylaluminum sesquichloride (EASC) for "prereduction" of $TiCl_4$ in hexane (eq 4.5). EASC reduces the oxidation state of titanium and $TiCl_3$ precipitates as the β (brown) form. Reduction is believed to proceed through an unstable alkylated Ti^{+4} species (eq 4.5) which decomposes to Ti^{+3} (eq 4.6). Lower oxidation states ($Ti+^2$) may also be formed. These reactions are exothermic and very fast.

$$TiCl_4 + (C_2H_5)_3Al_2Cl_3 \rightarrow Cl_3TiC_2H_5 + 2\ C_2H_5AlCl_2 \tag{4.5}$$

$$Cl_3TiC_2H_5 \rightarrow TiCl_3 \downarrow + \tfrac{1}{2}\,C_2H_4 + \tfrac{1}{2}\,C_2H_6 \tag{4.6}$$

By-product ethylaluminum dichloride (EADC) is soluble in hexane, but is a poor cocatalyst. EADC must be removed (or converted to a more effective cocatalyst) before introduction of monomer. For example, ethylaluminum dichloride can be easily converted (as in eq 4.7) to diethylaluminum chloride by redistribution reaction with triethylaluminum (see reference 3 and literature cited therein for discussions of aluminum alkyl redistribution reactions).

$$C_2H_5AlCl_2 + (C_2H_5)_3Al \rightarrow 2(C_2H_5)_2AlCl \tag{4.7}$$

Aluminum alkyls are still used industrially for prereduction of transition metal compounds. However, far more is used in the role of cocatalyst, described in the next section.

4.2.2   Alkylating Agent for Creation of Active Centers

In this case, the aluminum alkyl is functioning as a cocatalyst, sometimes also called an "activator." Titanium alkyls, believed to be active centers for polymerization, are created through transfer of an alkyl from aluminum to titanium, known as "alkylation." Molar ratios of cocatalyst to transition metal (Al/Ti) are typically ~30 for commercial polyethylene processes using Ziegler-Natta catalysts (lower ratios are used for polypropylene). The vast majority of aluminum alkyls sold into the polyethylene industry today is for use as cocatalysts. With TEAL, the most widely used cocatalyst, alkylation proceeds as in eq 4.8:

The titanium alkyl active center may be associated with (or stabilized by) an aluminum alkyl (see discussion on p. 41–42 and eq. 3.6).

4.2.3   Scavenger of Catalyst Poisons

In commercial polyethylene operations, poisons may enter the process as trace (ppm) contaminants in ethylene, comonomer, hydrogen (CTA), nitrogen (used as inert gas), solvents and other raw materials. These poisons reduce catalyst activity. Most damaging are oxygen and water. However, CO_2, CO, alcohols, acetylenics, dienes, sulfur-containing compounds and other protic and polar contaminants can also lower catalyst performance. With the exception of CO, aluminum alkyls react with contaminants converting them to alkylaluminum derivatives that are less harmful to catalyst performance. Illustrative reactions of contaminants with triethylaluminum are provided in eq 4.9–4.11:

$$(C_2H_5)_3Al + \tfrac{1}{2} O_2 \rightarrow (C_2H_5)_2AlOC_2H_5 \qquad (4.9)$$

$$2\,(C_2H_5)_3Al + H_2O \rightarrow (C_2H_5)_2Al\text{-}O\text{-}Al(C_2H_5)_2 + 2\,C_2H_6 \uparrow \qquad (4.10)$$

$$(C_2H_5)_3Al + CO_2 \longrightarrow (C_2H_5)_2AlO\overset{\overset{\displaystyle O}{\|}}{C}C_2H_5 \qquad (4.11)$$

Products from eq 4.9–4.11 may undergo additional reactions to form other alkyl-aluminum compounds. Since CO is not reactive with aluminum alkyls, it must be removed by conversion to CO_2 in fixed beds.

As mentioned in section 4.2.2, Ziegler-Natta catalyst systems used in the poly-ethylene industry typically employ high ratios of Al to transition metal in the polymerization reactor. Ratios of ~30 are common. Hence, there is a large excess of aluminum alkyl to achieve the roles depicted in sections 4.2.2 and 4.2.3 and to scavenge poisons.

4.2.4   Chain Transfer Agent

Chain transfer for Ziegler-Natta polyethylene catalysts is accomplished largely with hydrogen, as previously shown (see eq 3.7 in Chapter 3). However, at very high Al/Ti ratios, molecular weight of the polymer can be marginally lowered by chain transfer to aluminum. This occurs by ligand exchange between tita-nium and aluminum, previously illustrated in eq 3.10 of Chapter 3.

4.3   Magnesium Alkyls in Ziegler-Natta Catalysts

In the late 1960s, it was discovered that inorganic magnesium compounds, espe-cially $MgCl_2$ (see section 3.5) are excellent supports for Ziegler-Natta polyolefin catalysts. No doubt sparked by this discovery, chemists in the polyolefins industry began exploring use of magnesium alkyls in catalyst synthesis. The most well-known magnesium alkyls are alkylmagnesium halides. These reagents were discovered by Victor Grignard in 1905, and have become known as Grignard reagents, written "RMgX," where R is usually a simple alkyl and X a chloride ligand. The closely related dialkylmagnesium compounds (R_2Mg) received little attention relative to Grignard reagents which are ubiquitous in organic synthesis. In large part, this was due to the excellence of Grignard reagents for formation of new C-C bonds.

Though Grignard reagents were discovered more than a century ago, they remain key synthetic tools in the repertoire of the modern chemist. Indeed, Morrison and Boyd in their classic text describe the Grignard reagent as "one of the most useful and versatile reagents known to the organic chemist" (18). However, for many of the polyethylene catalyst preparations using magnesium alkyls emerging in the 1970s, Grignard reagents were not the material of choice. This was due primarily to the catalyst deactivating influence of the Lewis basic solvents used in their preparation, usually diethyl ether or tetrahydrofuran. Catalyst development chemists needed *hydrocarbon-soluble* magnesium alkyls. Unfortunately, there were few practical options available at the time, since the hydrocarbon solubility was poor for most Grignard reagents and R_2Mg known in the art of the 1970s.

Using a variety of approaches (3), hydrocarbon-soluble R_2Mg were discovered, including the so-called "unsymmetrical" dialkylmagnesium compounds (RMgR', where R and R' are C_2 to n-C_8 alkyl groups). In fact, RMgR' compounds are today the most important of the commercially available dialkylmagnesiums used in manufacture of polyethylene:

- n-butylethylmagnesium (BEM),
- n-butyl-n-octylmagnesium (BOM or BOMAG®), and
- "dibutylmagnesium" (DBM)

DBM contains both n-butyl and *sec*-butyl groups (ratio ~1.5) and may be regarded as a "mixture" of di-n-butylmagnesium (DNBM) and di-*sec*-butylmagnesium (see section 4.4). Likewise, BEM and BOM may be viewed as "mixtures" of DNBM with diethylmagnesium and di-n-octylmagnesium, respectively. However, because dynamic alkyl ligand exchange occurs, these compositions are not true mixtures. Neat (solvent-free) RMgR' are either viscous liquids or amorphous white solids and are commercially available only in hydrocarbon solution (usually heptane). RMgR' compounds have been commercially available since the mid-1970s. Though BEM was discovered in 1978 (19), it remains today the most important dialkylmagnesium compound for production of polyethylene catalysts.

In a 1985 review article, it was declared that dialkylmagnesium compounds are of "limited practical importance" (2). Though this statement may have been accurate at that time, it is certainly not true today. In the early 21st century, the value of polyethylene produced from R_2Mg-derived Ziegler-Natta catalysts contributes substantially to the "practical importance" of these materials. Globally, millions of tons of polyethylene are manufactured each year using such catalysts. Dialkylmagnesium compounds are used in Ziegler-Natta catalyst synthesis in two fundamentally different ways discussed in sections 4.3.2 and 4.3.3 below.

BEM, BOM and DBM are offered in slightly different formulations by several producers in the US and Europe. These companies supply R_2Mg formulations to polyolefin producers worldwide. However, no manufacturer offers all products. Major suppliers are:

- Akzo Nobel (formerly Texas Alkyls, Inc.)
- Albemarle (formerly Ethyl Corp.)
- Chemtura (formerly Crompton, Witco and Schering)
- FMC (formerly Lithium Corp. of America)

4.3.1 R_2Mg for Production of Supports

Reaction of dialkylmagnesium compounds with selected chlorinated compounds produces finely divided $MgCl_2$ that can be used as a support for polyethylene catalysts. Other reagents may be used to produce different inorganic magnesium compounds, also suitable as supports. Examples are shown in Figure 4.1. Treatment of these products with transition metal compounds results in a supported "precatalyst." Typically, the transition metal is subsequently reduced by reaction with an aluminum alkyl and the solid catalyst isolated. The solid catalyst and cocatalyst (usually TEAL) may then be introduced to the polymerization reactor.

As discussed in section on p. 37, fine particles are usually not desirable for Ziegler-Natta catalysts. However, most polyethylene catalysts produced by the method described in this section are used in solution processes where psd and morphology are less important than in gas phase or slurry processes (see Chapter 7).

4.3.2 R_2Mg as Reducing Agent

Reaction of R_2Mg with a transition metal compound produces a reduced transition metal composition co-precipitated with an inorganic magnesium compound. In this respect, dialkylmagnesium compounds are functioning in much the same way as aluminum alkyls described in section 4.2.2. As before, additional aluminum alkyl cocatalyst must be introduced in the polymerization reactor to alkylate the transition metal and create active centers.

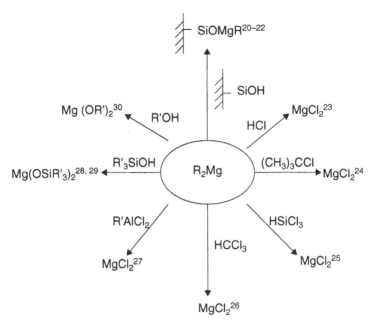

Figure 4.1 Reactions of dialkylmagnesium compounds to produce inorganic magnesium compounds useful as supports for Ziegler-Natta catalysts.

An additional point should be made before concluding discussion of the role of dialkylmagnesium compounds in Ziegler-Natta catalyst systems. There are occasional citations in the literature suggesting that dialkylmagnesium compounds are used as "cocatalysts." These references are probably using "cocatalysts" in the broadest context. That is, they are merely suggesting that dialkylmagnesium compounds are amongst components employed in the overall catalyst synthesis. As reducing agents and agents for production of supports, dialkylmagnesium compounds are highly effective. But as cocatalysts are defined in section 4.2.3, dialkylmagnesium compounds are very poor and may even completely deactivate some Ziegler-Natta catalysts. This may occur because of deactivation of active centers caused by strong coordination of the magnesium alkyl or perhaps overreduction of the transition metal.

4.4 Lithium Alkyls

Though not as important as aluminum and magnesium alkyls, lithium alkyls are produced in million pound per year quantities, but have only an indirect role in the production of polyethylene. The primary merchant application for lithium alkyls is in anionic polymerization of dienes to produce synthetic elastomers such as polybutadiene. Polybutadiene has many applications in the automotive industry (tires and radiator hoses, for example).

The largest volume commercial alkyllithium compound is *n*-butyllithium, but significant quantities of *sec*-butyl- and *tert*-butyllithium are also produced. Production of *n*-butyllithium is achieved by reaction of *n*-butyl chloride with lithium metal as in eq 4.12:

$$CH_3CH_2CH_2CH_2Cl + 2\,Li \rightarrow CH_3CH_2CH_2CH_2Li + LiCl \downarrow \quad (4.12)$$

Analogous reactions are used to manufacture *sec*-butyllithium and *tert*-butyllithium using 2-chlorobutane (*sec*-butyl chloride) and 2 chloro-2-methylpropane (*tert*-butyl chloride), respectively, as shown in eq 4.13 and 4.14:

$$
\begin{array}{ccc}
\overset{\displaystyle Cl}{\underset{\displaystyle |}{}} & & \overset{\displaystyle Li}{\underset{\displaystyle |}{}} \\
CH_3CH_2CHCH_3 + 2\,Li & \longrightarrow & CH_3CH_2CHCH_3 + LiCl \downarrow \quad (4.13)
\end{array}
$$

$$
\begin{array}{ccc}
\overset{\displaystyle CH_3}{\underset{\displaystyle |}{}} & & \overset{\displaystyle CH_3}{\underset{\displaystyle |}{}} \\
CH_3C\text{-}Cl + 2\,Li & \longrightarrow & CH_3C\text{-}Li + LiCl \downarrow \quad (4.14) \\
\underset{\displaystyle |}{} & & \underset{\displaystyle |}{} \\
CH_3 & & CH_3
\end{array}
$$

Reactions are conducted in an aliphatic hydrocarbon, usually hexane or isopentane, in which the lithium compound is soluble. Products are sold as dilute solutions (typically 15–25%)

The indirect role of lithium alkyls in polyethylene production involves the use of *sec*-butyllithium (from eq 4.13) to manufacture "dibutylmagnesium" (DBM). Key synthesis reactions in aliphatic hydrocarbon solvents (*e.g.*, heptane) are illustrated in eq 4.15–4.17:

$$2\,CH_3CH_2CH_2CH_2Cl + 2\,Mg \rightarrow (CH_3CH_2CH_2CH_2)_2Mg \downarrow + MgCl_2\downarrow \quad (4.15)$$

di-*n*-butylmagnesium (insoluble)

$$
\begin{array}{ccc}
\overset{\displaystyle Li}{\underset{\displaystyle |}{}} & & CH_3CH_2CHCH_3 \\
& & | \\
2\,CH_3CH_2CHCH_3 + MgCl_2 & \longrightarrow & Mg \qquad + \quad 2\,LiCl \downarrow \;(4.16) \\
& & | \\
& & CH_3CH_2CHCH_3
\end{array}
$$

di-*sec*-butylmagnesium (soluble)

$$\underset{\text{DSBM (soluble)}}{\left.\begin{array}{c} CH_3CH_2CHCH_3 \\ | \\ Mg \\ | \\ CH_3CH_2CHCH_3 \end{array}\right.} + \underset{\text{DNBM (insoluble)}}{(CH_3CH_2CH_2CH_2)_2Mg \downarrow} \longrightarrow 2 \underset{\text{DBM (soluble)}}{\left\{\begin{array}{c} CH_3CH_2CHCH_3 \\ | \\ Mg \\ | \\ CH_3CH_2CH_2CH_2 \end{array}\right\}} \quad (4.17)$$

The product in eq 4.17 results from dynamic ligand exchange ("redistribution reaction") of DSBM and DNBM. $MgCl_2$ in eq 4.16 may be the insoluble by-product from synthesis of di-n-butylmagnesium in eq 4.15. This makes it possible to produce DBM in a "one-pot" synthesis. Though DBM in eq 4.17 is shown as if the n-butyl to sec-butyl ratio were ~1, in practice the ratio is about 1.5. This minimizes the amount of costly sec-butyllithium that is needed. DBM is used to produce supported transition metal catalysts (discussed in sections 4.3.2 and 4.3.3).

4.5   Organoboron Compounds

Organoboron compounds constitute a broad and rich area of organometallic chemistry and a detailed discussion is inappropriate for an introductory text on polyethylene. However, several organoboron compounds are crucial for selected polyethylene catalyst technologies. For example, arylboranes are used as cocatalysts for single site catalyst systems and will be discussed in Chapter 6 (see section 6.3.2). The purpose of this section is to introduce the trialkylborane that is a component of 3rd generation Phillips catalyst systems (Chapter 5) employed in industrial processes in for linear polyethylene.

The most important of the simple trialkylboranes is triethylborane (TEB). It has been produced commercially since the 1960s. At ambient temperatures, triethylborane is a clear, colorless liquid (bp 95 °C) that is pyrophoric, and burns with a green flame (3). However, unlike most aluminum and magnesium alkyls, TEB is monomeric and is virtually unreactive with water. It is produced commercially by reaction of diborane (B_2H_6) with ethylene or by alkylation of a trialkyl borate by triethylaluminum as illustrated in eq 4.18 and 4.19, respectively.

$$3\ (CH_2{=}CH_2) + \tfrac{1}{2}\ B_2H_6 \rightarrow (CH_3CH_2)_3B \qquad (4.18)$$

$$3\ (CH_3CH_2)_3Al + (RO)_3B \rightarrow (CH_3CH_2)_3B + 3\ (CH_3CH_2)_2AlOR \quad (4.19)$$

4.6   Zinc Alkyls

Zinc alkyls have been known to chemical science since the mid-nineteenth century and were among the first organometallic compounds produced and characterized. Sir Edward Frankland, an English chemist and pioneer in organometallic chemistry, synthesized diethylzinc (DEZ) from zinc metal and ethyl iodide (3). Remarkably, more than century and a half after its discovery, diethylzinc remains today an important industrial metal alkyl. Though quantities are substantially smaller than those of aluminum alkyls, diethylzinc has several niche applications in polyethylene processes.

Diethylzinc, produced in commercial quantities (tons) since the late 1960s, exists as a clear, colorless liquid at ambient temperatures. It is monomeric and distillable (bp ~ 117 °C). DEZ is pyrophoric and reacts vigorously with water, though reaction is not as difficult to control as trialkylaluminum reactions with water. DEZ is unreactive with CO_2.

Diethylzinc is employed in several ways in polyethylene production. Its earliest application was as a chain transfer agent for molecular weight control (31, 32). Today, chain transfer in Ziegler-Natta catalyst systems is achieved chiefly by hydrogenolysis (previously discussed in connection with the mechanism of chain termination in section 3.7). Use of diethylzinc for molecular weight control for polyethylene is no longer significant.

Chromium catalysts are notorious for the difficulty of initiating polymerization after a "turnaround." When process equipment is taken out of service for maintenance, the interior of reactors may be exposed to ambient air. This introduces oxygen and water, severe poisons for chromium catalysts. Even after inert gas (nitrogen) is re-introduced after maintenance activities are completed, trace amounts of poisons adhere to interior surfaces. Diethylzinc is aggressively reactive with water and oxygen and is used to scavenge these poisons from polymerization reactors. When reactors are started up again, polymerization initiates more readily.

Uses of diethylzinc as a scavenger of poisons and as a chain transfer agent are relatively low-volume applications. However, a new industrial application of diethylzinc was recently disclosed (33, 34) that may have greater long range significance. Diethylzinc is used in production of Dow's INFUSE® block copolymers of ethylene and octene-1. A mixed single site catalyst system involving hafnium and zirconium is used. The mechanism is termed "chain shuttling." It is believed to occur by transfer of polymeric chains between transition metals through the intermediacy of diethylzinc. This is consistent with the propensity of diethylzinc to function as a chain transfer agent.

4.7   Safety and Handling of Metal Alkyls

Polyethylene producers that use Ziegler-Natta, single site and selected chromium catalysts are required to handle metal alkyls on a large-scale (in some cases, tons per year). As previously noted, many metal alkyls are pyrophoric, *i.e.*, they ignite spontaneously upon exposure to air. Most are also explosively reactive with water. Polyethylene manufacturers must routinely deal with these hazardous chemicals. Despite an abundance of resources and training aids from metal alkyl suppliers, accidents occur and severe injuries and even death have resulted. Clearly, safety and handling of metal alkyls must be a high priority.

Safety measures that should be taken while handling metal alkyls are best viewed as a cascading series of defenses against mishap. The first line of defense must be the person at the "front line", *i.e.*, the person making a transfer of a metal alkyl. That person must be thoroughly trained and practiced on safe procedures for transferring metal alkyls. He or she must not take short cuts or circumvent precautions. Transfer lines must be properly constructed and must include purge-capability following transfer. In preparation for a transfer, the equipment must be closely checked using inert gas pressure (nitrogen is commonly used) to insure that it is leak-free.

While handling metal alkyls, the last line of defense is to insure that the appropriate personal protective equipment (PPE) is worn. Major commercial suppliers recommend a variety of personal protective equipment for large-scale handling of metal alkyls including a flame-resistant hood, fire resistant coat and leggings, and impervious gloves (leather gloves or felt-lined PVC are most often used). Technical bulletins covering procedures for large-scale handling and transfer of metal alkyls are available from the major producers listed in section 4.2.1.

For small-scale laboratory transfers, recommended personal protective equipment includes a *minimum* of a full face shield and fire retardant lab coat. A recent tragic occurrence which resulted in the death of a university laboratory student (35) while handling *tert*-butyllithium would likely not have happened had the person been wearing proper personal protective equipment. A detailed discussion of techniques for safe laboratory transfers of metal alkyls is available (36).

Irrespective of the quantity of metal alkyl to be transferred, the worker wearing proper personal protective equipment will likely avoid severe injury in the event of an accident resulting from an unsafe act or equipment failure.

Proper personal protective equipment must be worn for large-scale transfers of metal alkyls. (Photo courtesy of Akzo Nobel Polymer Chemicals).

Unlike organic peroxides (see section 2.3 in Chapter 2), metal alkyls do not in general decompose violently if subjected to heat. Thermal decomposition of metal alkyls usually occurs in a slow, non-hazardous manner. However, there are a few exceptions. For example, diethylzinc and trimethylaluminum can decompose violently at elevated temperatures. *Extreme caution must be exercised to insure that these neat products are not subjected to high temperatures.*

Thermal stability of aluminum alkyls is highly dependent upon the ligands bonded to aluminum. For example, triethylaluminum is stable up to about 120 °C (37) and its initial mode of decomposition is by β-hydride elimination to generate diethyl-aluminum hydride and ethylene (eq 4.20). Diethylaluminum hydride may decompose further to aluminum, hydrogen and two additional equivalents of ethylene (eq 4.21). The overall equation for decomposition is shown in eq 4.22.

$$(CH_3CH_2)_3Al \rightarrow (CH_3CH_2)_2AlH + CH_2=CH_2 \qquad (4.20)$$

$$(CH_3CH_2)_2AlH \rightarrow Al + \tfrac{1}{2}\,H_2 + 2\,CH_2=CH_2 \qquad (4.21)$$

$$(CH_3CH_2)_3Al \rightarrow Al + 1\tfrac{1}{2}\,H_2 + 3\,CH_2=CH_2 \qquad (4.22)$$

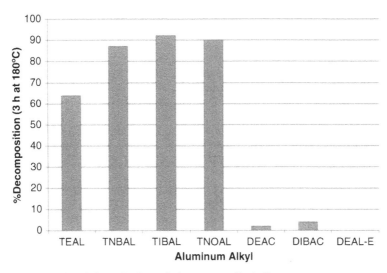

Figure 4.2 Thermal stability of selected aluminum alkyls.[37]

If, however, one of the ethyl groups in triethylaluminum is replaced by an ethoxide ligand, the resultant molecule is dramatically more stable thermally. For example, diethylaluminum ethoxide is stable up to at least 192 °C (37). A comparison of the percent decomposition of selected aluminum alkyls over 3 hours at 180 °C is shown in Figure 4.2. Clearly, trialkylaluminum compounds are much less thermally stable than derivatives containing an electron-rich heteroatom, *e.g.*, chlorine or oxygen.

For most industrial polyethylene processes (slurry and gas phase), thermal stability of the cocatalyst is not a factor since most operate in the temperature range 80–110 °C. However, solution processes operate at high enough temperatures where thermal decomposition of the cocatalyst could become a factor. Fortunately, residence times are typically short in solution processes.

References

1. JR Zietz, Jr., *Ullman's Encyclopedia of Industrial Chemistry*, Vol. A1, VCH Verlagschellshaft, Weinheim, FRG, 1985, p 543.
2. F Bickelhaupt and O Akkerman, *Ullman's Encyclopedia of Industrial Chemistry*, Vol. A15, VCH Verlagschellshaft, Weinheim, FRG, p 626, 1985.
3. DB Malpass, LW Fannin and JJ Ligi, *Kirk-Othmer Encyclopedia of Chemical Technology*, John Wiley and Sons, New York, Third Edition, Volume 16, p 559, 1981; see also DB Malpass, *Handbook of Transition Metal Catalysts*, R Hoff and R Mathers (editors), Wiley, Chapter 1, 2010.
4. FR Hartley and S Patai, *The Chemistry of the Metal Carbon Bond*, John Wiley and Sons, New York: Vol 1, *The Structure, Preparation, Thermochemistry and Characterization of Organometallic Compounds*, 1983; Vol 2, *The Nature and Cleavage of Metal-Carbon Bonds*,

1984; Vol 3, *Carbon-Carbon Bond Formation Using Organometallic Compounds*, 1985; Vol 4, *The Use of Organometallic Compounds in Organic Synthesis*, 1987.

5. JJ Eisch, *Comprehensive Organometallic Chemistry*, Vol 1, p 555, 1982.
6. JJ Eisch, *Comprehensive Organometallic Chemistry II*, Vol 1, p 431, 1995.
7. WE Lindsell, *Comprehensive Organometallic Chemistry*, Vol 1, p 155, 1982.
8. WE Lindsell, *Comprehensive Organometallic Chemistry II*, Vol 1, p 57, 1995.
9. JR Zietz, Jr., GC Robinson and KL Lindsay, *Comprehensive Organometallic Chemistry*, Vol 7, p 368, 1982.
10. MJ Krause, F Orlandi, AT Saurage and JR Zietz, Jr, *Ullman's Encyclopedia of Industrial Chemistry*, Wiley-VCH Verlag GmbH & Co. KGaA, Weinheim, 2005.
11. K Ziegler, *Organometallic Chemistry*, (ACS Monograph 147, H. Zeiss, editor), Reinhold, NY, p 194, 1960.
12. AH Tullo, *Chemical & Engineering News*, p 14, November 2, 2009.
13. D.B. Malpass, *Methylaluminum Compounds*, Society of Plastics Engineers (SPE). The International Polyolefins Conference, Houston, TX, February 25–28, 2001.
14. LM Allen, RO Hagerty and RO Mohring, US Patent 4,732,882, March 22, 1988.
15. Isoprenylaluminum is produced by reaction of isoprene (2-methyl-1,3-butadiene) with TIBAL or DIBAL-H; see JJ Ligi and DB Malpass, *Encyclopedia of Chemical Processing and Design*, Marcel Dekker, New York, Vol 3, p 32, 1977.
16. J Ehlers and J Walter, US Patent 5,587,440, December 24, 1996.
17. DB Malpass, US Patent 4,593,010, June 3, 1986.
18. RT Morrison and RN Boyd, *Organic Chemistry*, 6th edition, Prentice Hall, p 99, 1992.
19. LW Fannin and DB Malpass, US Patent 4,127,507, Nov. 28, 1978.
20. AD Caunt, PD Gavens and J McMeeking, US Patent 4,385, 161, May 24, 1983.
21. L van de Leemput, European Pat. Applic. 72,591, Feb. 23, 1983.
22. M Shida, TJ Pullukat and RE Hoff, US Patent 4,263,171, April 21, 1981 and US Patent 4,383,096, May 10, 1983.
23. DF Birkelbach and GW Knight, US Patent 4,198,315, April 15, 1980.
24. JC Bailly and J Collomb, US Patent 4,487,846, June 16, 1983.
25. H Sakurai, H Morita, T Ikegami and S Tsuyama, US Patent 4,159,965, July 3, 1979.
26. D Kurz, US Patent 4,366,298, Dec 28, 1981 and US Patent 4,368,306, Jan 11, 1983.
27. KP Wagner, US Patent 4,186,107, Jan 28, 1980.
28. CT Berge, MP Mack and CM Starks, US Patent 4,374,755, Feb 22, 1983.
29. WJ Heilman and RA Kemp, US Patent 4,525,557, June 25, 1985.
30. DE Gessell, US Patent 4,244,838, Jan 13, 1981.
31. EJ Vandenberg and BC Repka, High Polymers, (ed. CE Schildknecht and I Skeist), John Wiley & Sons, 1977, 29, p 370.
32. BA Krentsel, YV Kissin, VJ Kleiner and LL Stotskaya, Polymers and Copolymers of Higher α-Olefins, Hanser/Gardner Publications, Inc., Cincinnati, OH, p 46 (1997).
33. K Swogger, International Conference on Polyolefins, Society of Plastics Engineers, Houston, TX, February 25–28, 2007.
34. S Martin, International Conference on Polyolefins, Society of Plastics Engineers, Houston, TX, February 24–27, 2008.
35. JN Kemsley, Learning from UCLA, *Chemical & Engineering News*, p 29, August 3, 2009.
36. Anon., *Handling Air-sensitive Reagents*, Technical Bulletin AL-134, Aldrich Chemical Company, Inc., March, 1997.
37. G. Sakharovskaya, N. Korneev, N. Smirnov and A. Popov, *J. Gen. Chem. USSR*, (English edition), 1974, 44, 560.

5

Chromium Catalysts

5.1 Chromium Catalysts Supported on Metal Oxides

After World War II, US oil companies Phillips Petroleum and Standard Oil Company of Indiana assigned chemists to explore ways to convert olefins (ubiquitous by-products from various petroleum refining processes) to gasoline-type fuels or lubricants (1, 2). During this pursuit, transition metal catalysts supported on refractory oxides were discovered for polymerization of ethylene to linear polyethylene (3, 4). Among these were chromium catalysts supported on silica ("Phillips catalysts") discovered by Hogan and Banks and molybdenum catalysts supported on alumina ("Standard of Indiana catalysts") discovered by Zletz and coworkers. Chromium catalysts were aggressively promoted and licensed by Phillips (now Chevron Phillips Chemical Company) and have become among the most widely used catalysts for HDPE worldwide. During the dedication of their laboratory as a National Historic Chemical Landmark by the American Chemical Society in 1999, the innovative work by Hogan and Banks was termed a "company-maker" (1). Hogan and Banks were awarded the Perkin Medal by the Society of Chemical Industry in 1987. In contrast, Standard of Indiana (later to become Amoco and BP) pursued their catalysts rather passively and such catalysts have become mere footnotes in the historical development of commercial polyethylene. "Standard of Indiana" catalysts are today of limited industrial importance and will not be discussed further [for details, see McMillan (2) and Boor (3)].

The Phillips and Standard of Indiana discoveries predated by a substantial margin the work of Ziegler and Natta on transition-metal catalyzed olefin polymerizations in Europe. There is a considerable irony in the story of the Phillips catalyst and polypropylene patent rights. While it is true that the Phillips catalyst is today enormously important in manufacture of polyethylene (accounting for approximately a quarter of all polyethylene produced globally), it is unsatisfactory for commercial production of polypropylene. Nevertheless, one of the early experiments performed by Hogan and Banks involved propylene. They succeeded in producing a small quantity of crystalline polypropylene on June 5, 1951 with a Cr-on-silica catalyst (5), well before Natta and Ziegler. After litigation spanning more than 24 years, this experiment ultimately resulted in the composition-of-matter patent (US Patent 4,376,851) on "crystalline polypropylene" being awarded to Phillips in 1983.

Though Phillips catalysts are by far the most important supported chromium catalysts for polyethylene, there are other commercially important examples of such catalysts. These were developed primarily in the 1970s by the Union Carbide Corporation (6), now part of the Dow Chemical Company. UCC chromium catalysts for polyethylene will be discussed in section 5.4.

5.2   Basic Chemistry of Phillips Catalysts

A refractory oxide (usually silica) is essential to the performance of Phillips catalysts. Indeed, chromium compounds used in the production of Phillips catalysts are not stable at elevated temperatures and would decompose rapidly under the conditions typically employed for activation of Phillips catalysts (>500 °C). For example, CrO_3 decomposes to Cr_2O_3 and O_2 above 200 °C. However, when affixed ("chemisorbed") onto silica surfaces, supported Cr compositions are stable to temperatures at least up to 1000 °C (7).

Silicas suitable as supports for polyethylene catalysts are commercially available from several companies in a variety of catalyst grades. These typically have high surface areas (300–600 m^2/g) and large pore volumes (1–3 mL/g). Silicas are granular or spheroidal and are available in a range of average particle sizes, typically between about 40 and 150 microns. Particle size distribution must be controlled to avoid problems associated with overly large or small particles. As discussed previously, morphology and psd of the catalyst are replicated in the polymer (see section 3.5 and Figures 3.3 and 3.4). Key suppliers are:

- Grace-Davison
- INEOS Silicas (formerly Crosfields)
- PQ Corporation

Silica contains surface siloxane (Si-O-Si) and silanol (SiOH) groupings and relatively large amounts (4–7% by wt) of adsorbed water. These features are depicted schematically in Figure 5.1.

The chemistry of Phillips catalysts is intimately linked to interactions between inorganic chromium compounds and silica. Basic steps are summarized below:

- Treatment of silica with a chromium compound in aqueous solution. The chromium compound later reacts with surface silanol groups thereby affixing Cr to the surface of the support, either as the monochromate or dichromate, as illustrated in Figure 5.2.

- Removal of the solvent at ~100 °C.

- Calcination of chromium-containing silica at high temperature (typically 500–900 °C) to activate the catalyst. Activation is usually done in a fluidized bed using an oxidizing medium (air or oxygen) as the medium to insure that Cr remains largely in the +6 oxidation state. Cr content at completion of the activation step is typically about 1% (wt).

Chromium is present chiefly as the monochromate; however, some is chemisorbed as the dichromate. Though it is known that monochromate species are precursors to active centers for polymerization, it is possible that dichromate may also form active sites (7). However, monochromate is widely believed to be the dominant precursor and will be used in this discussion.

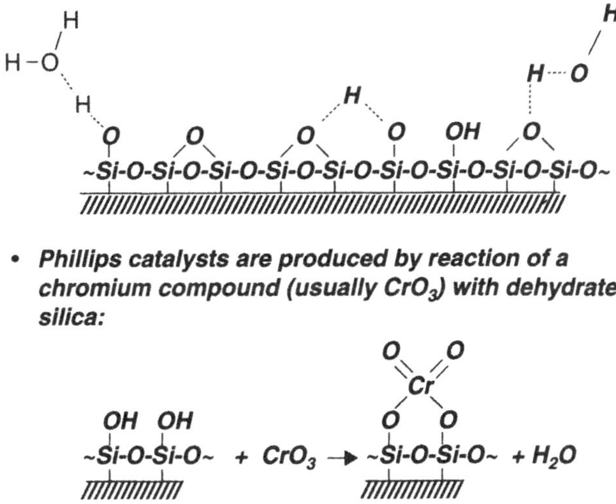

- *Phillips catalysts are produced by reaction of a chromium compound (usually CrO₃) with dehydrated silica:*

Figure 5.1 Silica surfaces contain high concentrations of hydrogen-bonded adsorbed water, siloxane (Si-O-Si) groupings and intramolecular hydrogen bonds via silanol (SiOH) groupings.

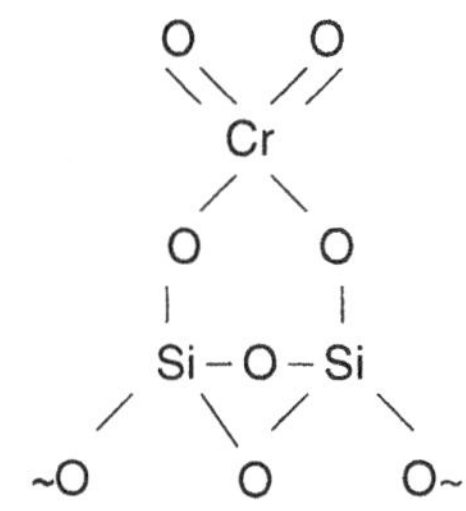

silyl monochromate

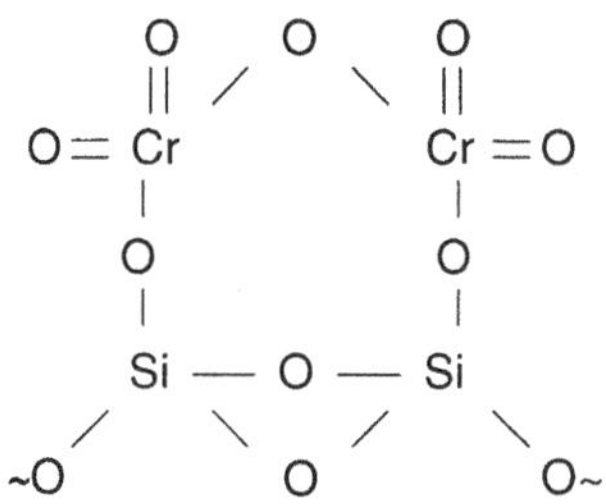

silyl dichromate

Figure 5.2 Surface chromate structures resulting from treatment of silica with CrO_3. Calcination at high temperatures in air (or O_2) insures that Cr remains primarily in the +6 oxidation state and results in Phillips catalyst for polyethylene. Final activation occurs in reactor through reactions with ethylene (see text for details).

After activation, the catalyst is introduced into the polymerization reactor as slurry in a saturated hydrocarbon such as isobutane. The precise mechanism of initiation is not known, but is believed to involve oxidation-reduction reactions between ethylene and chromium, resulting in formation of chromium (II) which is the precursor for the active center. Polymerization is initially slow, possibly because oxidation products coordinate with (and block) active centers. Consequently, standard Phillips catalysts typically exhibit an induction period. The typical kinetic profile for a Phillips catalyst is shown in curve C of Figure 3.1. If the catalyst is pre-reduced by carbon monoxide, the induction period is not observed. Unlike Ziegler-Natta and most single site catalysts, no cocatalyst is required for standard Phillips catalysts. Molecular weight distribution of the polymer is broad because of the variety of active centers.

Productivity of the Phillips catalysts is typically ~3 kg/g of catalyst (8). At the typical chromium loading of ~1%, residual chromium in the polymer is less than 5 ppm, and post-reactor treatment to remove catalyst residues is not necessary.

5.3   Generations of Phillips Catalysts

Since discovery in the early 1950s, a variety of modifications have been made to improve the Phillips catalyst. Beaulieu has identified four generations of Phillips catalysts (9), summarized in Table 5.1. These modifications are primarily intended to affect molecular weight and molecular weight distribution of the polymer, rather than catalyst activity.

A limitation of the 1st generation Phillips catalyst discovered by Hogan and Banks was its inability to produce high melt index (low molecular weight) resins (10). The 1st generation catalyst has poor hydrogen response for molecular weight control (11). Most commonly, marginal molecular weight control is achieved by polymerization temperatures between about 90 and 110 °C.

Second generation Phillips catalysts involve use of titanium compounds that modify the surface chemistry of the support and enables production of polyethylene with higher MI (lower MW) (12). Titanium tetraisopropoxide, also known as tetraisopropyl titanate (TIPT), is the most commonly used modifier for these catalysts. Hexavalent chromium titanate species are probably formed on the surface as shown in Figure 5.3 (13). Catalyst surfaces contain a diversity of active sites and molecular weight distribution of the polymer is broader than that from 1st generation catalysts.

Third generation Phillips catalysts use triethylborane (TEB) as an adjuvant. In these systems, triethylborane is sometimes called a "cocatalyst." However, it

Table 5.1 Principal features of generations of Phillips catalysts (B. Beaulieu, M. McDaniel and P. DesLauriers, Society of Plastics Engineers, Houston, TX, February 27, 2005).

Generation	Period	Features
1st	1955	Chromium on silica; after induction period produces high MW polymer with relatively broad MWD.
2nd	1975	TIPT used to modify support surface; results in lower polymer MW.
3rd	1983	TEB used as "cocatalyst"; broadens MWD of polymer by increasing low MW fraction
4th	1990s	Aluminum phosphate, alumina or silica-alumina used as support; PE with bimodal MWD may be produced; shows good hydrogen response, unlike other generations of Phillips catalysts.

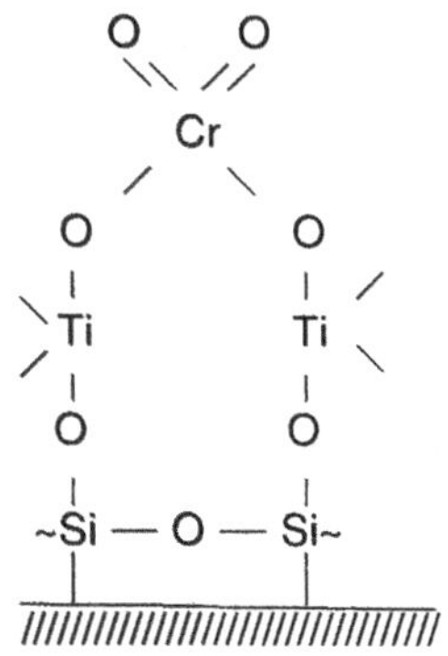

Figure 5.3 Structure of hexavalent chromium titanate species formed in 2nd generation Phillips catalyst. Catalyst produces polyethylene with higher MI (lower molecular weight) and broader MWD than chromium on silica alone.

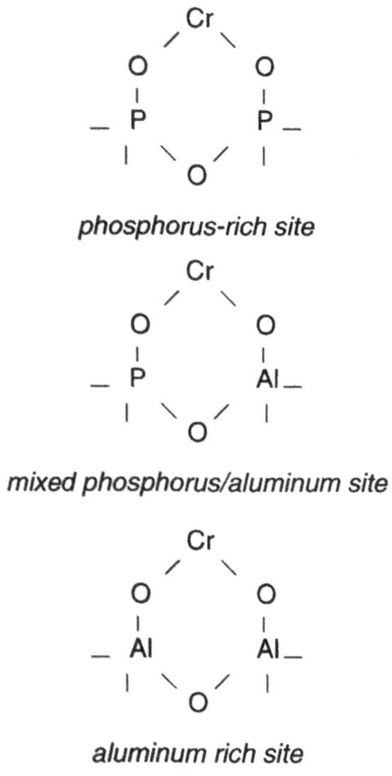

phosphorus-rich site

mixed phosphorus/aluminum site

aluminum rich site

Figure 5.4 Structures proposed for 4th generation Phillips catalyst. Because of diversity of active centers, catalyst produces polyethylene with broadest MWD ($M_w/M_n > 50$) relative to any other single catalyst in commercial use (7).

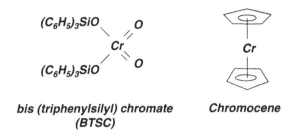

bis (triphenylsilyl) chromate
(BTSC)   **Chromocene**

Figure 5.5 Compounds used to produce supported chromium catalysts developed by Union Carbide for use in gas phase processes for LLDPE and HDPE. Catalysts must be supported, usually on silica, for optimal performance. Chromocene catalyst is used without a cocatalyst; BTSC uses diethylaluminum ethoxide as cocatalyst.

is important to note that triethylborane appears to function differently than aluminum alkyl cocatalysts in Ziegler-Natta systems (as described in section 4.2.3) and is used in much lower concentrations. Triethylborane broadens molecular weight distribution of the polyethylene and also generates hexene-1 *in situ* which lowers density and results in production of linear low density polyethylene.

Fourth generation Phillips catalysts employ aluminum phosphate ($AlPO_4$) and alumina as supports in place of silica. Unlike the 1st generation Phillips catalyst, Cr on $AlPO_4$ is responsive to hydrogen, allowing a wide range of MWs to be produced by adjusting hydrogen concentration in the reactor. $AlPO_4$ is isoelectric with SiO_2 and produces a variety of active centers, as shown in Figure 5.4. Each active center has a different reactivity with components in the reactor (monomer, comonomer, hydrogen, etc.). This results in a very broad molecular weight distribution ($M_w/M_n > 50$), claimed to be the highest polydispersity produced commercially from a single catalyst (7).

5.4   Union Carbide Chromium Catalysts

Supported chromium catalysts were developed by Union Carbide Corporation in the 1970s using different chromium precursors than are used in standard Phillips catalysts (6, 11). The most important of these are based on chromocene and bis(triphenylsilyl)chromate, depicted in Figure 5.5. These catalysts are used in the Unipol® gas phase process for LLDPE and HDPE and are different from standard Phillips catalysts in several respects:

- Kinetic profiles are different.
- Induction period is eliminated or minimized.

- Chromocene catalyst has excellent hydrogen response for molecular weight control.

5.5   Mechanism of Polymerization with Supported Chromium Catalysts

The initial step in the mechanism of ethylene polymerization using Phillips catalysts is believed to occur by way of an oxidation-reduction reaction between Cr (VI) and ethylene as depicted in eq 5.1. This generates Cr (II) and vacant coordination sites. As mentioned above, polymerization may be initially slow because of sluggish reduction or desorption of the oxidation by-products which can coordinate with (and block) active centers.

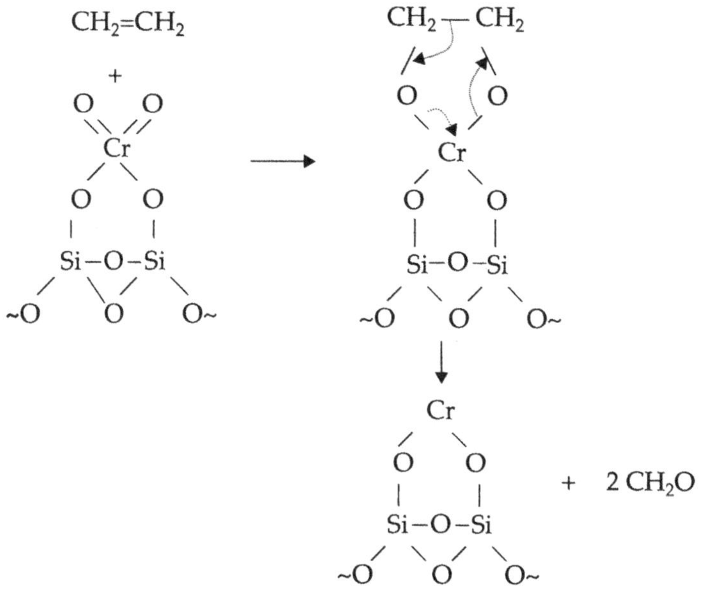

$$
\text{(5.1)}
$$

It is speculated that a chromium hydride or alkyl species is formed, but the exact mechanism for its formation is not known, though approximately six decades have elapsed since the basic discoveries of Hogan and Banks. The hydride (or alkyl) initially forms a π–complex with ethylene. The π–complex collapses to an alkylchromium moiety that serves as the active center for the polymerization (eq 5.2).

$$\begin{array}{c} Cr \\ \diagup \quad \diagdown \\ O \qquad O \\ | \qquad | \\ Si\text{-}O\text{-}Si \\ \diagup \diagdown \diagup \diagdown \\ \sim O \quad O \quad O\sim \end{array} \xrightarrow{\ ?\ } \begin{array}{c} \quad \searrow\!\!\diagdown \quad H \\ Cr \diagup \\ \diagup \quad \diagdown \\ O \qquad O \\ | \qquad | \\ Si\text{-}O\text{-}Si \\ \diagup \diagdown \diagup \diagdown \\ \sim O \quad O \quad O\sim \end{array}$$

$$CH_2=CH_2 \searrow$$

$$\begin{array}{c} /\!/ \diagdown \quad H \\ \quad Cr \diagup \\ \diagup \quad \diagdown \\ O \qquad O \\ | \qquad | \\ Si\text{-}O\text{-}Si \\ \diagup \diagdown \diagup \diagdown \\ \sim O \quad O \quad O\sim \end{array} \longrightarrow \begin{array}{c} \quad \square \quad CH_2CH_3 \\ \quad Cr \diagup \\ \diagup \quad \diagdown \\ O \qquad O \\ | \qquad | \\ Si\text{-}O\text{-}Si \\ \diagup \diagdown \diagup \diagdown \\ \sim O \quad O \quad O\sim \end{array} \qquad (5.2)$$

$$n\, CH_2=CH_2 \searrow$$

$$\begin{array}{c} \quad \square \quad (CH_2CH_2)_nCH_2CH_3 \\ \quad Cr \diagup \\ \diagup \quad \diagdown \\ O \qquad O \\ | \qquad | \\ Si\text{-}O\text{-}Si \\ \diagup \diagdown \diagup \diagdown \\ \sim O \quad O \quad O\sim \end{array}$$

Chain termination occurs primarily by β-elimination with hydride transfer to chromium and by β-elimination with hydride transfer to monomer. These terminations are analogous to those previously shown for Ziegler-Natta polymerizations (see eq 3.8 and 3.9 in Chapter 3). In some cases, supported chromium catalysts, $e.g.$, chromocene on SiO_2 and Cr on $AlPO_4$, are responsive to hydrogen for molecular weight control and chain transfer to hydrogen can sometimes dominate in these systems.

Polymerizations of ethylene using chromium catalysts have been reviewed in depth by Hogan and McDaniel (14–19).

References

1. K. MacDermott, *Chemical & Engineering News*, p 49, November 29, 1999.
2. FM McMillan, *The Chain Straighteners*, MacMillan Press, London, p 71, 1979.
3. J Boor, Jr., *Ziegler-Natta Catalysts and Polymerizations*, Academic Press, Inc., p 280, 1979.
4. RB Seymour, *Advances in Polyolefins*, Plenum Press, New York, p 6, 1987.
5. JP Hogan and RL Banks, *History of Polyolefins*, (RB Seymour and T Cheng, editors), D. Reidel Publishing Co., Dordrecht, Holland, p 105, 1986.
6. FJ Karol, *Encyclopedia of Polymer Science and Technology*, Supp Vol 1, p 120, 1976.
7. PJ DesLauriers, M McDaniel, DC Rohlfing, RK Krishnaswamy, SJ Secora, PL Maeger, EA Benham, AR Wolfe, A. Sukhadia and B. Beaulieu, *International Conference on Polyolefins*, Society of Plastics Engineers, Houston, TX, February 25–28, 2007.
8. M Smith, *Handbook of Petrochemicals Production Processes*, McGraw-Hill, p 14.35, 2005.
9. B Beaulieu, M McDaniel and P DesLauriers, *International Conference on Polyolefins*, Society of Plastics Engineers, Houston, TX, February 27, 2005.
10. HL Hsieh, MP McDaniel, JL Martin, PD Smith and DR Fahey, *Advances in Polyolefins*, Plenum Press, New York, 153, 1987.
11. FJ Karol, BE Wagner, IJ Levine, GL Goeke and A Noshay, *Advances in Polyolefins*, Plenum Press, New York, 339, 1987.
12. MP McDaniel, MB Welch and MJ Dreiling, *J. Catal.*, 1983, *82*, 118.
13. TJ Pullukat, RE Hoff and M Shida, *J. Polym. Sci, Polymer Chemistry Ed.*, 1980, *18*, 2857.
14. J. P. Hogan, *J. Polymer Science, Part A-1*, 1970, *8*, 2637.
15. J. P. Hogan, *Applied Industrial Catalysis*, 1983, *1*, 149, Ch 6, BE Leach (editor), Academic, New York.
16. MP McDaniel, *Handbook of Heterogeneous Catalysis*, 2nd ed; 2007, chapter 15.1, G Ertl, H Knozinger, F Schuth, J Weitkamp, (editors), Wiley-VCH Verlag, Weinheim, Germany.
17. MP McDaniel, *Handbook of Heterogeneous Catalysis*, 1st ed, 1997, G Ertl, H. Knozinger and J. Weitkamp, (editors), VCH Verlagsgesellschaft, Weinheim, Vol 5, 2400.
18. E Benham and M McDaniel in *Kirk-Othmer Concise Encyclopedia of Chemical Technology*, 5th ed, A Seidel (editor), John Wiley & Sons, Inc., Hoboken, 2007, 590.
19. MP McDaniel, *Handbook of Transition Metal Catalysts*, R Hoff and R Mathers (editors), Wiley, 2010.

6

Single Site Catalysts

6.1 Introduction

In several ways, single site catalysts are antithetical to Ziegler-Natta catalysts. SSC are based upon highly purified transition metal compounds, while most Ziegler-Natta catalysts are rather ill-defined heterogeneous mixtures. Cocatalysts for Ziegler-Natta polymerizations are well-characterized, pure compounds, but the most commonly used cocatalysts for single site catalysts (methylaluminoxanes) are relatively impure and are poorly understood. The active centers for single site catalysts are believed to be cationic, while active centers for Ziegler-Natta catalysts are thought to involve neutral octahedral complexes having open coordination sites. Ziegler-Natta catalysts have a multiplicity of active sites that polymerize ethylene in slightly different ways, resulting in polyethylene with typical polydispersities of 4–6. Single site catalysts have essentially one type of active center ("single site") and produce polyethylene with *very* narrow MWD (polydispersities of 2–3). Contrasting features of single site catalysts and Ziegler-Natta catalysts are summarized in Table 6.1.

Most single site catalysts are homogeneous (1). Consequently, psd and morphological control of catalyst and polymer particles are not possible. For solution processes, this is of little concern. However, for slurry/suspension and

Table 6.1 Contrasting characteristics of single-site and Ziegler-Natta catalysts.

	Single Site Catalysts	Ziegler-Natta Catalysts
Typical Catalyst:	metallocenes of Zr and Ti	amorphous solid containing Ti
Purity of catalyst:	High	Low
Typical cocatalyst:	MAO	TEAL
Purity of cocatalyst:	Low	High
Active centers:	single site	multiple sites
Polymer Mw/Mn:	2–3	4–6

gas phase processes, psd and morphology may be critical. Hlatky has reviewed supported single site catalysts for olefin polymerization (1).

There are two types of single site catalysts. The most well-known are based on metallocenes. Non-metallocene types are a relatively recent development and most are based upon chelated compounds of late transition metals, primarily Pd, Ni and Fe. Each single site catalyst type is addressed below.

Before concluding this section, a fact that puts single site catalysts into proper perspective should be reiterated: Single site catalysts, while important technologically, contribute less than 4% of the global industrial production of polyethylene as of this writing (2). This may well change significantly in the future. However, the likely scenario is one wherein Ziegler-Natta, Phillips and free radical polymerizations remain the dominant methods for production of polyethylene for many decades to come.

6.2 Metallocene Single Site Catalysts

Metallocenes are π-bonded organometallics (3, 4) in which a metal is "sandwiched" between aromatic ligands, such as dicyclopentadienyl or indenyl groups. In metallocenes such as ferrocene, the cyclopentadienyl rings are parallel, but others have "bent sandwich" structures, as depicted in Figure 6.1. The controlled geometry catalyst (CGC) used in Dow's Insite® single site catalyst technology for polyethylene is an example of a "half sandwich" metallocene, depicted in Figure 6.2. Examples of metallocenes used to produce stereoregular polypropylene are shown in Figure 6.3. Metallocenes combined with methylaluminoxanes or fluoroarylboranes are the most widely used single site catalyst systems. Though not all are effective, selected metallocenes are highly active catalysts for ethylene polymerization, delivering activities >10^6 g PE/ g Met-atm C_2H_4-h, where "Met" is usually Zr or Ti.

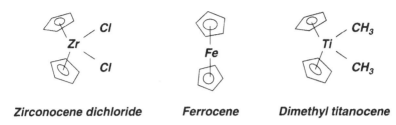

Zirconocene dichloride   **Ferrocene**   **Dimethyl titanocene**

Figure 6.1 Structure of simple metallocenes. Ferrocene was the first metallocene (discovered in 1951), but the correct π-bonded structure was not identified until 1952 (JP Collman, LS Hegebus, JR Norton and RG Finke, *Principles and Applications of Organotransition Metal Chemistry*, University Science Books, Sausalito, CA, p 9 (1987).

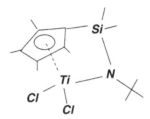

Figure 6.2 Structure of a CGC useful for solution polymerization of ethylene (JC Stevens, 11th Int'l Congress on Catalysts 40th Anniv., *Studies in Surface Science and Catalysis*, Vol. 101, p 11, 1996; see also K Swogger, International Conference on Polyolefins, Society of Plastics Engineers, Houston, TX, February 25–28, 2007).

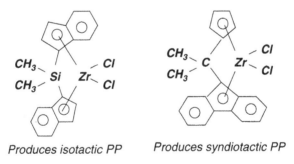

Produces isotactic PP   *Produces syndiotactic PP*

Figure 6.3 Structures of metallocene single site catalysts used to produce stereoregular polypropylene (M. P. Stevens, *Polymer Chemistry*, 3rd Edition, Oxford University Press, p 246, 1999).

Though metallocenes have been known since 1951 (5), it was not until the work of Kaminsky, Sinn and coworkers (6, 7) in the mid- and late-1970s that the enormous potential of metallocene single site catalysts was realized. The key discovery was the dramatic increase in activity resulting from use of methylaluminoxanes in place of diethylaluminum chloride and other conventional cocatalysts. Commercial use of metallocene single site catalysts began in the early 1990s.

Stevens (8) has described several advantages that Dow's constrained geometry catalyst has over other metallocenes used for ethylene polymerization:

- CGC has excellent high temperature stability and is capable of producing ethylene/octene-1 copolymers with $M_n \geq 5 \times 10^5$, even under the extreme conditions typical of solution processes.

- CGC has good hydrogen response, allowing a range of polymer molecular weights to be achieved. Polymer molecular weight may also be controlled in part by selection of process temperature. Higher temperatures afford polymer with lower molecular weight.

- CGC also has excellent reactivity with α-olefin comonomers. The latter attribute enables production of copolymers containing large amounts of uniformly distributed α-olefin (VLDPE). Further, by a mechanism wherein a long chain α-olefin is eliminated and subsequently incorporated as a "comonomer", a small amount of long chain branching is introduced into the polymer.

6.2.1 Non-metallocene Single Site Catalysts

Single site catalysts that are not derived from metallocenes were discovered in the 1990s. These catalysts are based primarily on chelated late transition metals, especially Pd, Ni and Fe (9, 10). An exemplary structure is shown in Figure 6.4. They offer many of the same advantages of metallocene single site catalysts, but are potentially less costly and are less oxophilic than metallocenes of early transition metals. Lower oxophilicity translates into greater compatibility with functional groups and ultimately the capability to produce copolymers of ethylene with polar comonomers. For example, vinyl acetate might be used to produce a predominantly *linear* version of EVA. The microstructure of such a copolymer will be vastly different from high pressure EVA, which is highly branched, and improved properties are anticipated (11). Non-metallocene single site catalysts

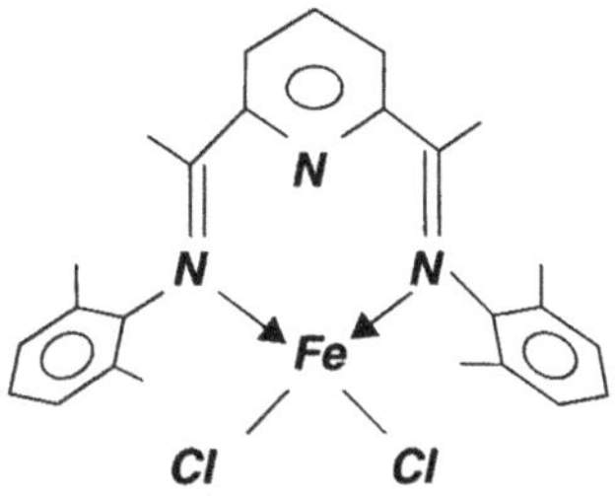

Figure 6.4 Non-metallocene single site catalysts based on chelated late transition metals are illustrated here with an iron catalyst (See B. Small, M. Brookhart and A. Bennett, *J. Am. Chem. Soc.*, 1998, 120, 4049; see also S. Ittel, L. Johnson and M. Brookhart, *Chem. Rev.* 2000, 100, 1169).

may also perform well with a broader range of cocatalysts (12). Indeed, in some cases, it is possible to eliminate expensive cocatalysts altogether (12).

Owing to a mechanism called "chain-walking," certain non-metallocene single site catalysts induce chain branching (13). Both short chain and long chain branching may result from the chain-walking mechanism. In principle, this makes it possible to produce highly branched polyethylene without use of comonomer.

The mechanism of chain walking has been discussed in a review (13). Briefly, the active center is able to migrate ("walk") down the polymer chain through a series of eliminations (involving β-agostic interactions) and re-additions, in some cases, even past tertiary carbon atoms. Migration of the active center induces branching in the resultant polymer. Using palladium catalysts, hyperbranched polyethylenes were produced with densities as low as 0.85 g/cc, without use of α-olefin comonomers. Key steps are illustrated with a palladium catalyst below:

Two recent developments in non-metallocene single site catalysts for polyethylene are noteworthy:

- Swogger has described "pyridyl amine catalysts" (2) which are based on early transition metals (Zr and Hf). An example of a pyridyl amine catalyst is provided in Figure 6.5. When two such catalysts are combined, the dual catalyst system is capable of producing olefin block copolymers of ethylene and octene-1 called INFUSE® through a mechanism termed "chain shuttling." Diethylzinc (DEZ) is the agent that promotes chain shuttling between two

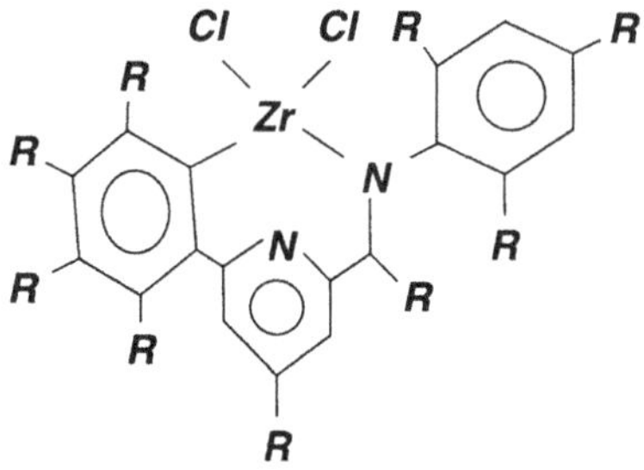

Figure 6.5 Structure of a pyridyl amine catalyst for production of ethylene/octene-1 block copolymers, where R is the same or different alkyls. (K Swogger, International Conference on Polyolefins, Society of Plastics Engineers, Houston, TX, February 25–28, 2007.)

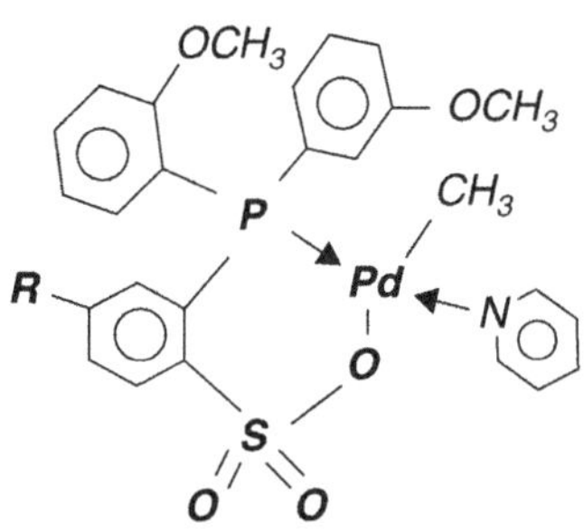

Figure 6.6 Structure of a single-site catalyst described by Goodall. Catalyst is capable of copolymerizing ethylene with polar comonomers without cocatalysts (BL Goodall, NT Allen, DM Conner, TC Kirk, LH McIntosh III and H Shen, International Conference on Polyolefins, Society of Plastics Engineers, Houston, TX, February 25–28, 2007).

such catalysts. (Diethylzinc is known to be an effective chain transfer agent for Ziegler-Natta catalysts. (14,15))

- Goodall (12) disclosed late transition metal catalysts that are highly active and are capable of copolymerizing ethylene with polar comonomers, such as acrylic acid and methyl acrylate. Moreover, Goodall catalysts do not require cocatalysts. An example of a Goodall catalyst is provided in Figure 6.6.

These developments and non-metallocene single site catalysts in general represent the next wave of innovation in polyolefin catalysis which should permit production of polyethylenes with unique properties at lower cost. They will complement, and perhaps even supplant, many of the metallocene single site catalysts commercialized in the 1990s.

6.3 Cocatalysts for Single Site Catalysts

As noted above, conventional aluminum alkyls are not effective cocatalysts for single site catalysts, probably because they are incapable of abstracting a ligand to generate the cationic active center (see section 6.4 on mechanism). Two main types of organometallic cocatalysts have been developed for use with single site catalysts. These are methylaluminoxanes (MAO) and arylboranes, both discussed below.

6.3.1   Methylaluminoxanes

Most commercially available methylaluminoxanes are produced by careful reaction of water with trimethylaluminum (TMAL) in toluene. Reaction must be closely controlled to avoid what renowned organometallic chemist John Eisch called "a life threatening pyrotechnic spectacle" (16). Unfortunately, there have been explosions and injuries reported during MAO preparations. Water must be introduced at low temperature and in forms that moderate the potentially violent reaction. For example, water has been introduced as hydrated salts, ice shavings or atomized spray. Even with these precautions, explosive reactions have occurred. The overall reaction is given in eq 6.1.

$$x(CH_3)_3Al + xH_2O \xrightarrow{\text{toluene}} \sim (CH_3AlO)_x \sim +2x\,CH_4\uparrow \qquad (6.1)$$

Yields are often low in lab preps (usually <60%). The product is called methyl-aluminoxane (MAO) or, less commonly, polymethylaluminoxane (PMAO). MAO is an ill-defined, complex composition, virtually insoluble in aliphatic hydrocarbons. MAO is typically supplied as a toluene solution containing ~13% Al, which corresponds to ~28% concentration of MAO.

Strides have been made in industrial-scale production of MAO. Process improvements afford greatly improved yields. This has been achieved by use of alternative reactants and/or continuous processes, employing highly dilute solutions, low ratios of water to TMAL and recycle of intermediate streams (17).

As isolated from toluene solution, neat MAO is an amorphous, friable white solid containing 43–44% Al (theory 46.5%). Like most commercially available aluminum alkyls, it is pyrophoric and explosively reactive with water. Freshly prepared MAO solutions form gels within a few days when stored at ambient temperatures (>20 °C). However, lower storage temperatures (0–5 °C) delay gel formation. Consequently, manufacturers store and transport MAO solutions in refrigerated containers. Commercially available MAO contains residual TMAL (15–30%), called "free TMAL" or "active aluminum." The literature is contradictory on the influence of free TMAL on activity of single site catalysts; both reductions and increases have been reported (18–20). Perhaps the most important drawback of methylaluminoxane is its cost, which is substantially higher than conventional aluminum alkyls. Despite these untoward aspects, methylaluminoxane remains the most widely used cocatalyst for industrial single site catalysts.

Other alkylaluminoxanes, *e.g.*, isobutylaluminoxane (IBAO), are also available, more easily produced and significantly less costly than methylaluminoxane.

However, these alternative aluminoxanes perform poorly as cocatalysts for single site catalysts. Preparation and properties of aluminoxanes have been extensively reviewed (20–22).

Published data on methylaluminoxane isolated from toluene have shown a wide range of molecular weights (300–3000 amu, primarily using cryoscopic methods). Possible reasons for the irreproducibility were proposed by Beard, et al. (23), who showed that cryoscopic molecular weight measurements of commercially available methylaluminoxane are influenced by several variables, such as process oils, residual toluene (solvent) and TMAL content. Beard reported "corrected" cryoscopic molecular weights of ~850, suggesting x in eq 16 to be ~15.

Alkylaluminoxanes have been shown to exist as highly associated oligomeric, cage or cluster structures (24, 25). Barron, et al., prepared t-butylaluminoxane (TBAO) by equimolar direct hydrolysis of tri-t-butylaluminum at −78 °C followed by thermolysis. TBAO was found to be primarily hexameric and nonameric, though some higher aggregates were also observed. (Use of a hydrated salt as the water source afforded different aggregates.) Isobutylaluminoxane, a commercially available alkylaluminoxane isomeric with TBAO, has been shown to have a cryoscopic MW of ~950 (26), in close agreement with nonameric association. Barron proposed that methylaluminoxane exists in cluster structures wherein aluminum is exclusively tetracoordinate. He further suggested that TMAL in commercial methylaluminoxane exists in two forms: dimeric TMAL ((CH_3)$_6$Al$_2$) and TMAL that is coordinated to oxygen atoms in the cluster (27). It is also possible for TMAL to associate with an Al-CH_3 group in the nonamer *via* electron deficient (three center-two electron) bonding (28, 29). A possible structure for adducts between TMAL and nonameric methylaluminoxane is illustrated below:

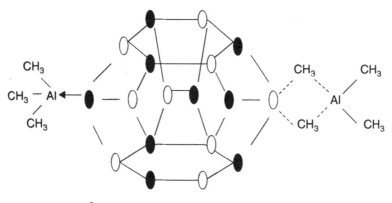

● = oxygen atoms in nonamer
◯ = aluminum atoms in nonamer
--- = electron deficient bonding between TMAL and an Al in nonamer
◀— = dative bond between oxygen in nonamer and Al in TMAL
Other methyl groups bonded to Al in nonamer omitted

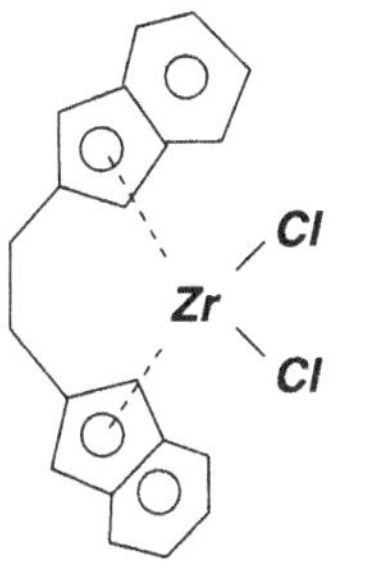

rac-ethylenebis(indenyl)zirconium dichloride (EBZ)

Figure 6.7 Structure of EBZ.

A nonhydrolytic method for production of methylaluminoxane suitable for single site catalysts has been reported (30–32). This alternative synthesis avoids altogether the hazardous reaction of TMAL with water and affords essentially quantitative recovery of aluminum values. Because the product provides higher activity in a standard ethylene polymerization test using *rac*-ethylenebis(indenyl) zirconium dichloride (EBZ, see Figure 6.7), it was dubbed PMAO-IP (from poly-methylaluminoxane-improved performance). Though many precursors may be used, the simplest method involves reaction of CO_2 with TMAL to form an intermediate. Subsequent thermolysis produces PMAO-IP. The detailed chemistry is complex and involves evolution of methane and other hydrocarbons, including products resulting from Friedel-Crafts reaction with toluene. A simplified equation is shown below (eq 6.2).

$$2\ (CH_3)_3Al + CO_2 \xrightarrow{\text{toluene}} (CH_3)_2AlOCOAl(CH_3)_2$$

with CH_3 substituents indicated on the central Al atoms

$$\longrightarrow\ \sim (CH_3AlO)_x \sim + CH_4 \text{ and other hydrocarbons} \qquad (6.2)$$

PMAO-IP contains much lower "free TMAL" than hydrolytic methylaluminoxane, which may explain the higher activity with selected single site catalysts. However, performance of PMAO-IP does not extend across the entire range of single site catalysts and it cannot be considered a "drop-in" replacement for standard methylaluminoxane.

Modified methylaluminoxanes (MMAO) have also been offered commercially since the early 1990s. MMAO (32) is a generic term encompassing all products wherein some of the methyl groups are replaced by other alkyl groups. The most commonly used modifiers are isobutyl and *n*-octyl groups.

Most modified methylaluminoxanes are prepared by reaction with water (eq 6.3). There are several formulations of MMAO (differentiated by a suffix, e.g., "MMAO-3A"), each with different composition and properties. One commercially available MMAO is produced by the nonhydrolytic

$$x \, R_3Al + x \, H_2O \rightarrow \sim (RAlO)_x\sim + 2x \, RH\uparrow \qquad (6.3)$$

method described above. All modified methylaluminoxanes contain $\geq$65% methyl groups and, as such, remain predominantly methylaluminoxane. Indeed, one MMAO formulation contains ~95% methyl groups.

Modified methylaluminoxanes exhibit much improved storage stability and several are highly soluble in aliphatic hydrocarbons. (Manufacturers of polyethylene prefer to avoid toluene because of toxicity concerns, especially if resins are destined for food contact.) Most importantly, because yields are higher, modified methylaluminoxane formulations are less costly than MAO. However, since modified methylaluminoxanes contain other types of alkylaluminoxanes, they do not match the performance of conventional methylaluminoxane in some single site catalyst systems. Consequently, modified methylaluminoxanes should be considered niche cocatalysts for single site catalysts.

All commercially available methylaluminoxanes employ trimethylaluminum as the starting material. As previously mentioned on p. 47, TMAL must be manufactured by less efficient processes that involve reduction by metallic sodium (33). Consequently, trimethylaluminum is much more expensive than other R_3Al compounds (~ten-fold). This, coupled with low yields of methylaluminoxanes from hydrolysis of trimethylaluminum, translates to very high costs for methylaluminoxanes. Additionally, methylaluminoxanes must be used in huge excess in many single site catalyst systems, further increasing the cost. (For example, ratios of Al to transition metal ratios are often >1000 in solution.) These factors provided impetus to develop less costly cocatalysts for single site catalysts. Even though alternative versions (non-hydrolytic and modified methylaluminoxanes) are obtained in higher yields, and lower ratios of Al/transition metals (<200) may be used with supported single site catalysts, use of methylaluminoxanes remains very costly relative to conventional aluminum alkyls. The next section describes the most successful alternatives to methylaluminoxanes.

6.3.2 Arylboranes

Tris(pentafluorophenyl)borane, known as "FAB" (structure below), is the most common arylborane used as cocatalyst for single site catalysts. FAB is a strongly Lewis acidic, air-sensitive solid (T_m 126–131 °C) that is only slightly soluble in hydrocarbon solvents. The structure of FAB is given below.

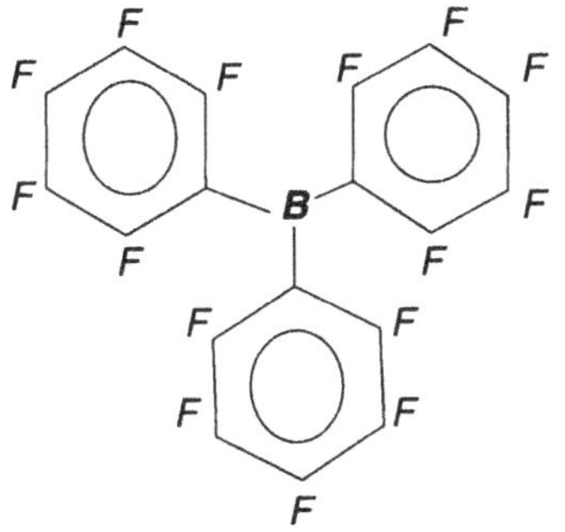

"FAB"

FAB may be derivatized further to produce cocatalysts that are even more Lewis acidic. For example, the following "ate" complexes (34) are also useful as cocatalysts:

$$Ph_3C^+(C_6F_5)_4B^-$$
trityl tetra(perfluorophenyl)borate

$$(CH_3)_2PhNH^+(C_6F_5)_4B^-$$
N,N-dimethylanilinium tetra(perfluorophenyl)borate

$$Li^+(C_6F_5)_4B^-$$
lithium tetra(perfluorophenyl)borate

The strong Lewis acidity of FAB and ate complexes enable them to abstract a ligand from the transition metal of the single site catalyst, creating a cation believed to be the active center for polymerization as illustrated in eq (6.4):

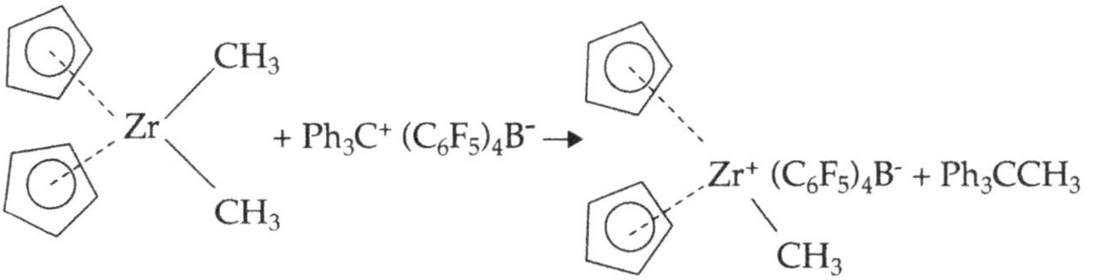

The main advantage of FAB and ate complexes is that they may be used in near stoichiometric amounts (35), unlike methylaluminoxanes which must be used in large excess for optimal results. Other arylboranes have also been used, some providing up to 20 times the activity of FAB in single site catalysts systems. Marks and Chen (20) have reviewed synthesis and properties of several of these alternative arylboranes.

6.3.3 Other Cocatalysts for Single Site Catalysts

Though methylaluminoxane, modified methylaluminoxanes and arylboranes/ borates are the cocatalysts most often used with single site catalysts, there are other compounds that function as cocatalysts. These include compositions such as Ph_3C^+, $Al(OC_6F_5)_4^-$. To date, however, these have not achieved significant usage in industry. **Caution:** *Aluminum compounds containing fluoroalkyl and fluoroaryl groups have been known to decompose violently when heated (20, 36).*

6.4   Mechanism of Polymerization with Single Site Catalysts

There are substantial differences between the mechanisms of polymerization with single site catalysts and the closely related Ziegler-Natta catalysts (37–42). Most notably, the active centers of single site catalysts are believed to be cationic. Currently, cocatalysts are used in all commercial processes using single site catalysts, but this may change in the not-too-distant future (see p. 76).

Active site generation is the essential first step and is illustrated in eq 6.5 with the relatively simple system of dimethylzirconocene and methylaluminoxane. Cocatalysts function as agents to create cationic active centers by ligand abstraction. It has been shown that the counterion in the ion pair resulting from abstraction must be weakly coordinated to the cationic active center. Cocatalysts such as methylaluminoxane and modified methylaluminoxanes, or the "free TMAL" contained therein, may also serve as alkylating agents and scavengers of poisons as they do in Ziegler-Natta catalyst systems (see sections 4.2.3 and 4.2.4). Propagation is shown in eq 6.6.

Initiation:

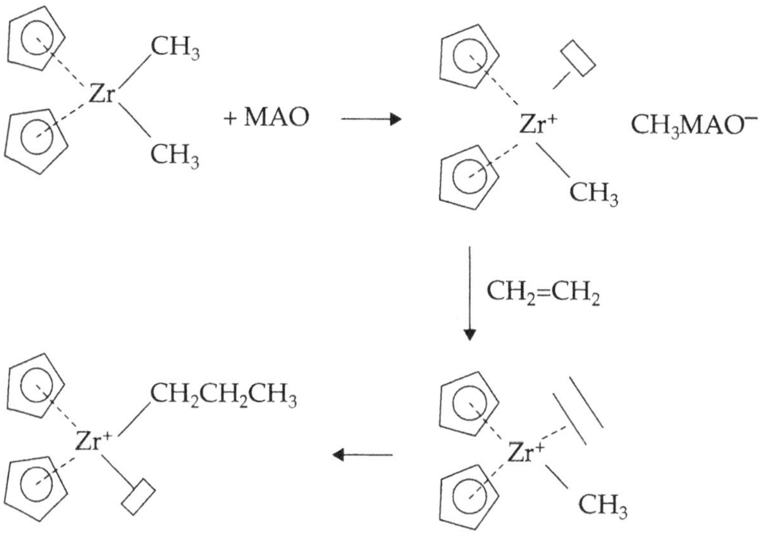

$$CH_2=CH_2 \tag{6.5}$$

Propagation:

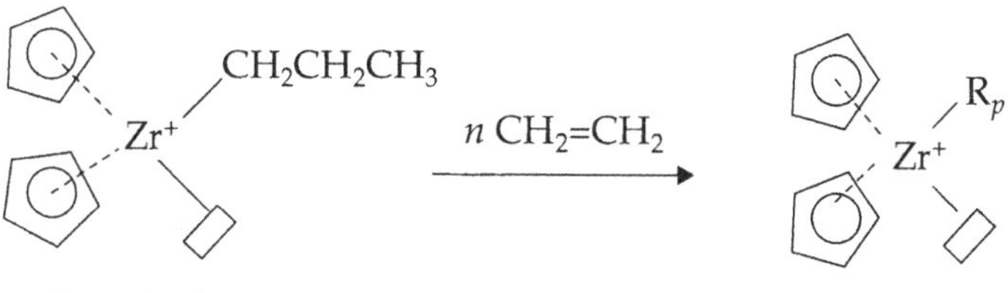

$$R_p = \text{-}(CH_2CH_2)_nCH_2CH_2CH_3$$

Termination:

As with Ziegler-Natta catalysts, termination may occur by several modes:

- chain transfer to hydrogen (hydrogenolysis)
- chain transfer by β-elimination with hydride transfer to transition metal
- chain transfer by β-elimination with hydride transfer to monomer

These mechanisms were previously illustrated in eq 3.7–3.9 of Chapter 3.

References

1. G Hlatky, *Chem. Rev.*, 2000, *100*, 1347.
2. K Swogger, *International Conference on Polyolefins*, Society of Plastics Engineers, Houston, TX, February 25–28, 2007.
3. JP Collman, LS Hegebus, JR Norton and RG Finke, *Principles and Applications of Organotransition Metal Chemistry*, University Science Books, Sausalito, CA, p 165, 1987.
4. RH Crabtree, *The Organometallic Chemistry of the Transition Metals*, Wiley-Interscience, New York, 3rd Ed., p 130, 2001.
5. JP Collman, LS Hegebus, JR Norton and RG Finke, *Principles and Applications of Organotransition Metal Chemistry*, University Science Books, Sausalito, CA, p 9, 1987.
6. H Sinn and W Kaminsky, *Adv. Organomet. Chem.*, 1980, *18*, 99.
7. H Sinn, W Kaminsky, HJ Wolmer and R Woldt, *Angew. Chem. Int. Ed. Engl.*, 1980, *19*, 390.
8. JC Stevens, *11th Int'l Congress on Catalysts-40th Anniversary, Studies in Surface Science and Catalysis*, Vol. 101, p 11, 1996.
9. SD Ittel, LK Johnson and M Brookhart, *Chem. Rev.*, 2000, *100*, 1169.
10. GJP Britovsek, VC Gibson, BS Kimberly, PJ Maddox, SJ McTavish, GA Solan, AJP White and DJ Williams, *Chem. Commun.*, 1998, 848.
11. LS Boffa and BM Novak, *Chem. Rev.*, 2000, *100*, 1479.
12. BL Goodall, NT Allen, DM Conner, TC Kirk, LH McIntosh III and H Shen, *International Conference on Polyolefins*, Society of Plastics Engineers, Houston, TX, February 25–28, 2007.
13. SD Ittel, LK Johnson and M Brookhart, *Chem. Rev.*, 2000, *100*, 1179.
14. EJ Vandenberg and BC Repka, *High Polymers*, (ed. CE Schildknecht and I Skeist), John Wiley & Sons, *29*, p 370, 1977.
15. BA Krentsel, YV Kissin, VJ Kleiner and LL Stotskaya, *Polymers and Copolymers of Higher α-Olefins*, Hanser/Gardner Publications, Inc., Cincinnati, OH, p 46, 1997.
16. JJ Eisch, *Comprehensive Organometallic Chemistry II*, Vol 1, p 451, 1995.
17. JK Roberg and EA Burt, US Patent 5,663,394, September 2, 1997.
18. I Tritto, C Mealares, MC Sacchi and P Locatelli, *Macromol. Chem. Phys.*, 1997, *198*, 3963.
19. SS Reddy, K Radhakrishnan and S Sivaram, *Polymer Bulletin*, 1996, *36*, 165.
20. EY-X Chen and TJ Marks, *Chem. Rev.*, 2000, *100*, 1395.
21. S Pasynkiewicz, *Polyhedron* 1990, *9*, 429.
22. SS Reddy and S. Sivaram, *Prog. Polym. Sci.*, 1995, *20*, 309.
23. WR Beard, DR Blevins, DW Inhoff, B Kneale and LS Simeral, *International Polyolefin Conference*, The Institute of Materials, London, Nov. 1997.

24. MR Mason, JM Smith and AR Barron, *J. Am. Chem. Soc.* 1993, *115*, 4971.
25. CJ Harlan, SG Bott, AR Barron, *J. Am. Chem. Soc.* 1995, *117*, 6465.
26. DB Malpass, *Properties of Aluminoxanes from Akzo Nobel*, Akzo Nobel Polymer Chemicals product pamphlet MA 03.324.01, January, 2003.
27. AR Barron, *Organometallics*, 1995, *14*, 3581.
28. JP Collman, LS Hegebus, JR Norton and RG Finke, *Principles and Applications of Organotransition Metal Chemistry*, University Science Books, Sausalito, CA, p 100, 1987.
29. K Ziegler, *Organometallic Chemistry*, (ACS Monograph 147, H. Zeiss, editor), Reinhold, NY, p 207, 1960.
30. GM Smith, JS Rogers and DB Malpass, *Proceedings of the 5th International Congress on Metallocene Polymers*, Düsseldorf, Germany, organized by Schotland Business Research, Inc., Skilman, NJ, March 31–April 1, 1998.
31. GM Smith, JS Rogers and DB Malpass, *Proceedings of MetCon '98*, organized by The Catalyst Group, Spring House, PA, June 10–11, 1998.
32. GM Smith, SW Palmaka, JS Rogers and DB Malpass, US Patent 5,381,109, Nov. 3, 1998.
33. DB Malpass, *Methylaluminum Compounds*, Polyolefins 2001-The International Conference on Polyolefins, South Texas Section of SPE, Houston, TX, Feb. 27, 2001.
34. MB Smith and J March, *March's Advanced Organic Chemistry*, John Wiley & Sons, New York, 5th ed., p 339, 2001.
35. GG Hlatky, HW Turner and RR Eckman, *J Am Chem Soc*, 1989, *111*, 2728.
36. DB Malpass, *Chemical & Engineering News*, April 2, 1990, p 2.
37. MP Stevens, *Polymer Chemistry*, 3rd ed., Oxford University Press, New York, p 246, 1999.
38. HH Brintzinger, D Fischer, R. Mulhaupt, B. Rieger and RM Waymouth, *Angew. Chem. Int. Ed. Engl.*, *34*, 1134, 1995.
39. VK Gupta, S Satish and IS Bhardwaj, *Rev Macromol Chem Phys*, 1994, *C34* (3), 438.
40. PC Mohring and NJ Coville, *J Organomet Chem*, 1994, *479*, 1.
41. AK Kulshreshtha and S Talapatra, *Handbook of Polyolefins*, C. Vasile, editor, Marcel Dekker, New York, 1, 2000.
42. J-I Imuta and N Kashiwa, *Handbook of Polyolefins*, C. Vasile, editor, Marcel Dekker, New York, 71, 2000.

7

An Overview of Industrial Polyethylene Processes

7.1 Introduction

Principal process technologies employed in the manufacture of polyethylene are:

- high pressure autoclave
- high pressure tubular
- slurry (suspension)
- gas phase
- solution

Reactors used in ethylene polymerizations range from simple autoclaves and steel piping to continuous stirred tank reactors (CSTR) and vertical fluidized beds. Since the 1990s, a trend has emerged wherein combinations of processes are used with transition metal catalysts. These combinations allow manufacturers to produce polyethylene with bimodal or broadened molecular weight distributions (see section 7.6).

Conditions used in PE processes vary widely. Because the heat of polymerization for ethylene is quite high (variably reported to be between 22 and 26 kcal/mole), efficient heat removal is crucial for polyethylene processes. Selection of process must also accommodate catalyst features, such as its kinetic profile. Table 7.1

Table 7.1 Features of key industrial polyethylene processes*.

Company	Process Tradename	Product(s)	Process Type	Catalyst(s)	Comments
Basell	Spherilene	HDPE, MDPE, VLDPE	gas phase	Ziegler-Natta	Two reactors used in series
Basell	Hostalen	HDPE	slurry	Ziegler-Natta	Two reactors may be used in series or in parallel; bimodal HDPE available
Basell	Lupotech G	HDPE, MDPE	gas phase	supported Cr	Use of alkylaluminum alkoxide as cocatalyst gives higher activity and minimizes induction period
Basell	Lupotech T	LDPE, EVA	high pressure tubular	organic peroxides or air	Operates @ 2000–3100 bar
Borealis	Borstar	HDPE, MDPE, LLDPE	slurry loop and gas phase in series	Ziegler-Natta	Catalyst prepolymerized
ChevronPhillips		HDPE, MDPE, LLDPE	slurry	supported Cr, Ziegler-Natta, single site	So-called "particle form loop slurry process" used
Dow	Dowlex	LLDPE	solution	Ziegler-Natta	Octene-1 may used as comonomer
Dow	Insite	LLDPE, VLDPE	solution	controlled geometry catalyst (single site)	Octene-1 may be used as comonomer

Company	Process	Products	Process type	Catalyst	Comments
DSM/ Stamicarbon		LLDPE	solution	Ziegler-Natta	High pressure autoclave used in so-called "compact process"
ExxonMobil		LDPE, EVA	high pressure autoclave and tubular (separate processes)	organic peroxides	Autoclave operates @ about 1600 bar, tubular @ about 2800 bar
INEOS	Innovene	LLDPE	gas phase	Ziegler-Natta	Catalyst originally developed by Naphtachimie
Mitsui		HDPE, LLDPE	slurry	Ziegler-Natta	
Nova**	Sclairtech	HDPE, LLDPE, VLDPE	solution	Ziegler-Natta, single-site	Dual CSTR can be used in parallel or in series to produce broad range of products
Polimeri Europa		LLDPE, LDPE, EVA	high pressure autoclave and tubular (separate processes)	organic peroxides, Ziegler-Natta	High pressure processes operate @ 200–300 Mpa for LDPE and EVA (EAA and EMA also available); ZN catalysts operate @ 50–80 MPa for LLDPE.
Univation Technologies	Unipol	HDPE, MDPE, LLDPE	gas phase	Ziegler-Natta, supported Cr, single site	Bimodal PE available with proprietary catalyst system

* Information gleaned primarily from the *Handbook of Petrochemicals Production Processes*, McGraw-Hill, edited by R. A. Meyers, 2005.

** Process originally developed by Dupont Canada.

summarizes principal features of industrial processes used in the manufacture of polyethylene.

LDPE is produced solely in high pressure or autoclave processes using free radical initiators. Most often, organic peroxides are used as initiators, though other compounds that readily undergo homolytic scission to generate free radicals may also be used. Transition metal catalysts (Ziegler-Natta, Phillips and single site) are used to manufacture VLDPE, LLDPE, MDPE and HDPE in slurry, gas phase and solution processes. Analysis of the global demand for the various forms of polyethylene suggests that approximately 27% of the total polyethylene produced worldwide is manufactured by high pressure processes. (See Chapter 8 for more information on markets and volumes.)

As mentioned in Chapter 1, ethylene is always the more reactive olefin in systems used to produce copolymers involving α-olefins (LLDPE and VLDPE). An important process consideration for copolymerizations is the reactivity ratio. This ratio may be used to estimate proportions needed in reactor feeds that will achieve the target resin. However, fine tuning is often required to obtain the density or comonomer content desired. Reactivity ratios were discussed previously (Chapter 2) in the context of free radical polymerization of ethylene with polar comonomers. Reactivity ratios are also important in systems that employ transition metal catalysts for copolymerization of ethylene with α-olefins to produce LLDPE. Discussions of derivations and an extensive listing of reactivity ratios for ethylene and the commonly used α-olefins are provided by Krentsel, *et al.* (1).

Purity of ethylene from modern refineries may be as high as 99.95% (2) and can often be fed directly to industrial polymerization processes. However, purification of raw materials is usually required in polyethylene processes that employ transition metal catalysts. Impurities such as H_2O, O_2, CO, CO_2, acetylenics and sulfur compounds can be particularly damaging to transition metal catalyst performance, even at ppm levels. As discussed in Chapter 4, aluminum alkyls employed as cocatalysts in Ziegler Natta catalyst systems have a mitigating influence on poisons by converting them to alkylaluminum derivatives that are not as harmful to the transition metal catalyst. However, chromium catalysts do not typically employ metal alkyls and this can sometimes cause difficulty in initiating polymerization. Chromium catalysts often show an induction period. This is thought to be a consequence of the mechanism of the initiation step which is believed to generate an oxygen-containing carbon compound that coordinates with (and blocks) active centers (see section 5.5 and eq 22).

7.2   High Pressure Processes

Among industrial processes for production of polyethylene, free radical polymerizations are conducted under the most severe conditions, typically employing temperatures of >200 °C and pressures of 15,000 to 45,000 psig. Free radical polymerizations are conducted adiabatically in thick-walled autoclaves or tubing. At such high temperatures, polymerization of ethylene occurs in "solution" of polymer in excess monomer. Diluents (solvents) are not needed. Polyethylene particles precipitate from excess monomer when the reaction mix cools.

Except for the reactor zones, autoclave and tubular processes are very similar (3, 4). Peripherals in both cases are designed pre-reactor to ramp pressures and temperatures to very high levels and post-reactor to reduce temperatures and lower pressures to near ambient conditions to enable product isolation. Simplified process flow diagrams for the autoclave and tubular processes are shown in Figures 7.1 and 7.2, respectively.

ExxonMobil has produced LDPE by both high pressure processes since the late 1960s. Schuster and Kaus have described the workings and advantages of ExxonMobil high pressure processes for LDPE (3).

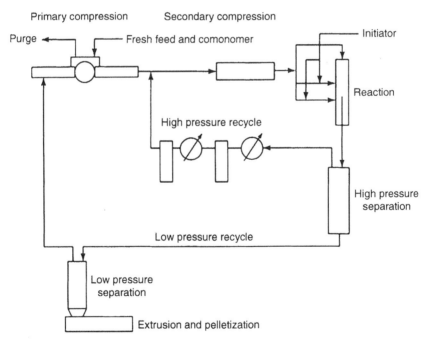

Figure 7.1 Schematic process flow diagram for autoclave high pressure process for production of low density polyethylene. (Reprinted with permission of John Wiley & Sons, Inc., *Kirk-Othmer Encyclopedia of Chemical Technology*, John Wiley and Sons, Inc., 6th edition, 2006).

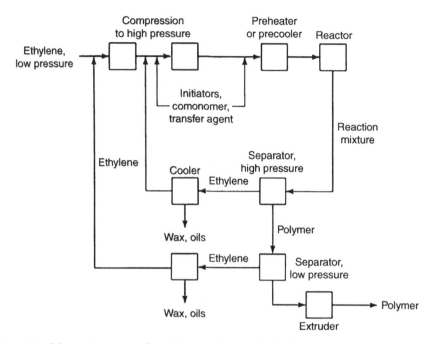

Figure 7.2 Schematic process flow diagram for tubular high pressure process for production of low density polyethylene. (Reprinted with permission of John Wiley & Sons, Inc., *Kirk-Othmer Encyclopedia of Chemical Technology*, John Wiley and Sons, Inc., 6th edition, 2006).

As discussed in Chapter 2 (sections 2.2 and 2.3), safety is a key consideration in high pressure processes. Handling organic peroxides presents potential hazards, but another serious concern is the possibility of ethylene decomposition within the reactor. High pressure processes must be designed to relieve pressure rapidly to prevent catastrophic explosions. Mirra of Polimeri Europa has described a reactor safety discharge system that quenches hot gases if an ethylene "decomp" takes place (4).

7.2.1   Autoclave Process

The original process for high pressure polyethylene was based on use of a high pressure autoclave and used air to introduce free radicals sufficient to initiate polymerization of ethylene. Principal features of the autoclave process are summarized in Table 7.2.

Use of air has been largely supplanted by organic peroxides (see section 2.3). Organic peroxides are injected at several points in the autoclave and initiates free radical polymerization by chemistry discussed in Chapter 2. Reactor residence

Table 7.2 Typical operating features of autoclave processes for LDPE.

Operating Temp:	• 180–300 °C
Operating Pressure:	• 15,000–30,000 psig
Features:	– continuous stirred tank reactors with agitators are used – multiple reactor zones typically used – polymerization takes place in "solution" – organic peroxides typically used as initiators – polymer has fewer branches but of somewhat longer length relative to tubular process

Table 7.3 Typical operating features of tubular processes for LDPE.

Operating Temp:	• 150–300 °C
Operating Pressure:	• 30,000–45,000 psig

Features:	– polymerization takes place in "solution"
	– tube is 1000–2000 m long and internal diameter of
	  25–50 mm (0.1–0.2 in)
	– organic peroxides typically used as initiators
	– polymer chain-length usually longer than product from
	  autoclave process, but with relatively short branches

times are very short (seconds or even fractions of a second). Excess ethylene is used to aid in heat removal (4). In general, purification of ethylene monomer is not necessary for high pressure processes. Unlike processes that use transition metal catalysts, high pressure processes are tolerant of trace amounts of water.

7.2.2   Tubular Process

The tubular process for LDPE may be considered to be a plug flow reactor. As in the autoclave process, organic peroxide initiator is injected at several points along the length of the tube. The tubes are typically 1000–2000 m long with an internal diameter of 25–50 mm (0.1–0.2 in). Product from the tubular process is typically higher in molecular weight and has more short chain branches than LDPE from the autoclave process. Key operating features of the tubular process are summarized in Table 7.3.

7.3   Slurry (Suspension) Process

Polymerizations may be conducted in diluents in which polyethylene is insoluble at the process temperature. Such processes are termed slurry (or suspension) processes.

Diluents must be inert toward the catalyst system and are usually saturated hydrocarbons such as propane, isobutane and hexane. Slurry processes typically operate at temperatures from about 80 to 110 °C and pressures of 200–500 psig. Polyethylene precipitates as formed resulting in a suspension of polymer in diluent. The catalysts most commonly used in slurry processes are chromium-on-silica or supported Ziegler-Natta catalysts.

Polymerizations that use supported chromium (Phillips) catalysts are conducted predominantly in slurry processes (though a small portion employs the gas phase process, see below). The historical development of the Phillips process has been expertly reviewed by Hogan (5, 6) and McDaniel (7–9). The slurry process originally developed by Phillips Petroleum (now Chevron Phillips) has been called the "particle form loop slurry process" and the "slurry loop reactor process" for production of HDPE and LLDPE (10). Hexene-1 is most often used as comonomer for LLDPE in the Phillips process. A simplified process flow diagram for the Phillips loop-slurry reactor process is shown in Figure 7.3 and key operating features are summarized in Table 7.4.

Another well-known slurry (suspension) process was developed by what was then Hoechst in Germany in the mid-1950s. Hoechst was the first licensee to use the catalyst and process developed by Karl Ziegler for producing low pressure

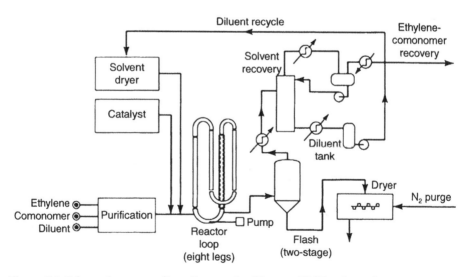

Figure 7.3 Schematic process flow diagram for Chevron Phillips loop slurry process for production of linear low density polyethylene. (Reprinted with permission of John Wiley & Sons, Inc., *Kirk-Othmer Encyclopedia of Chemical Technology*, John Wiley and Sons, Inc., 6th edition, 2006).

Table 7.4 Typical operating features of slurry/suspension processes for LLDPE and HDPE.

Operating Temp:	• 80–110 °C
Operating Pressure:	• 150–450 psig
Features:	– particles of growing polymer form as suspension in hydrocarbon diluent – catalyst residence time ~ 1 hour for Phillips loop slurry process – morphology and psd of catalyst are important – wide range of comonomers may be used

linear polyethylene in 1955. Hoechst was eventually absorbed into the company that is known today as LyondellBasell.

The Hoechst slurry process was improved over the years and has evolved into what is now called the Hostalen® process. Hostalen® is a slurry-cascade process that is capable of producing a wide range of molecular weight distributions of HDPE. The modern Hostalen® process employs 2 continuous stirred tank reactors that can be run in series or in parallel to produce unimodal and bimodal HDPE (11).

7.4   Gas Phase Process

Gas phase ethylene polymerizations are typically conducted in fluidized beds at pressures of 200–500 psig and temperatures of 80–110 °C. Gas phase processes for polyethylene were developed originally by Union Carbide (now Dow) and later by Naphtachimie (now INEOS). These processes are called the Unipol® and Innovene® processes, respectively. The predominant catalyst used in each process is of the supported Ziegler-Natta type, though the catalysts are produced by completely different chemistries. The Unipol® process is now licensed through Univation Technologies, a joint venture of Dow and ExxonMobil. Historically, the Unipol® process has dominated licenses for gas phase processes for linear polyethylene, but Innovene® has attracted a significant number of licensees in recent years.

Most of the polyethylene made in gas phase processes employs Ziegler-Natta catalysts. There are, however, a few instances where supported chromium and single site catalysts are used. A simplified process flow diagram for the Unipol® gas phase reactor process is shown in Figure 7.4. Principal operating features of the gas phase process are summarized in Table 7.5.

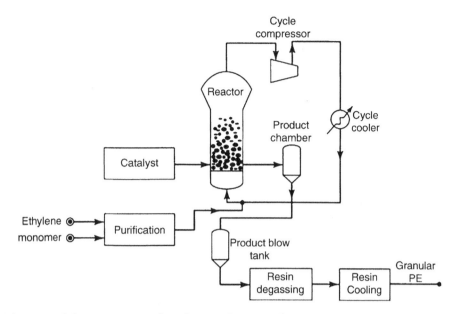

Figure 7.4 Schematic process flow diagram for Unipol® gas phase process for linear low density polyethylene. (Reprinted with permission of John Wiley & Sons, Inc., *Kirk-Othmer Encyclopedia of Chemical Technology*, John Wiley and Sons, Inc., 6th edition, 2006).

Table 7.5 Typical operating features of gas phase processes for HDPE and LLDPE.

Operating Temp:	• 80–110 °C
Operating Pressure:	• ~ 300 psig
Features:	– particles of growing polymer form in fluidized bed – catalyst residence time 2-4 hours – morphology and psd of catalyst are important – previously (pre-1990s) restricted in range of comonomer that could be used; because of emergence of "condensed" mode operation, a wide range of comonomers may now be used

In the 1990s, an improvement for the gas phase process was developed called "condensed mode" operation of Unipol® reactors (12). This technique greatly expanded capacity of gas phase reactors and product capability by making it more practical to use higher alpha-olefin comonomers such as octene-1.

7.5   Solution Process

In 1960, DuPont-Canada (now Nova) commercialized what has become known as the "solution process" using Ziegler-Natta catalysts based on titanium and vanadium compounds. DSM (Stamicarbon) and Dow also developed highly successful solution processes for polyethylene.

These processes employ mostly Ziegler-Natta catalysts. Solution processes operate at 160–220 °C and pressures of 500–5000 psig. Under such conditions, the polymer is dissolved in the solvent, typically cyclohexane or C_8 aliphatic hydrocarbons. Polymerization is homogeneous, occurring in solution at temperatures well above the melting range of polyethylene. A simplified process flow diagram for the Nova solution process is shown in Figure 7.5. Key operating features for the solution process are summarized in Table 7.6.

Dual continuous stirred tank reactors are used in the modern version of the DuPont-Canada/Nova solution process which is called "Advanced SCLAIRTECH" technology. Both Ziegler-Natta and single site catalysts may be used in Advanced SCLAIRTECH technology. The process is capable of producing a wide range of molecular weight distributions and densities ranging from VLDPE to HDPE. The SCLAIRTECH process for polyethylene has been described by Wiseman (13).

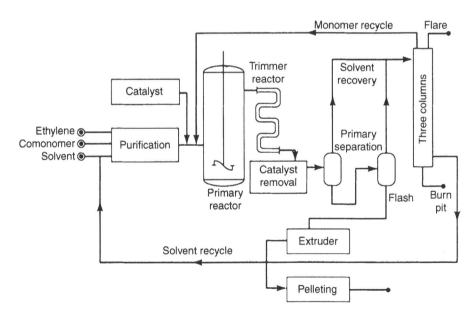

Figure 7.5 Schematic process flow diagram for DuPont-Canada (now Nova) solution process for production of polyethylene. (Reprinted with permission of John Wiley & Sons, Inc., *Kirk-Othmer Encyclopedia of Chemical Technology*, John Wiley and Sons, Inc., 6th edition, 2006).

Table 7.6 Typical operating features of solution processes for LLDPE and HDPE.

Operating Temp:	• 160–220 °C
Operating Pressure:	• 500–5000
Features:	– polymerization takes place in "solution" – catalyst residence time short (minutes) – catalyst and cocatalyst must show reasonably good high temperature stability – morphology and psd of catalyst are less important than in other processes – wide range of comonomers may be used

7.6 Combined Processes

As mentioned in section 7.1, technologies have been developed in recent years wherein combinations of processes are used to produce polyethylene. A case in point is the Borstar process developed by Borealis and started up in 1995. Borstar is capable of producing the entire range of polyethylenes from LLDPE to HDPE (14).

The Borstar process employs a small loop prepolymerization reactor (see section 3.6 for a discussion of the advantages of prepolymerization). Ziegler-Natta catalysts and triethylaluminum cocatalyst are commonly used but the process is capable of using single site catalysts (15).

Borstar also uses a large-scale loop slurry polymerization reactor and a gas phase polymerization reactor in series and is capable of producing bimodal polyethylene. The loop slurry reactor produces low molecular weight fractions and the gas phase reactor produces higher molecular weight product.

References

1. BA Krentsel, YV Kissin, VJ Kleiner and LL Stotskaya, *Polymers and Copolymers of Higher α-Olefins*, Hanser/Gardner Publications, Inc., Cincinnati, OH, p 245, 1997; see also YV Kissin, *Alkene Polymerizations with Transition Metal Catalysts*, Elsevier, The Netherlands, p 190, 2008.
2. K Weissermel and H.-J. Arpe, *Industrial Organic Chemistry*, Wiley-VCH, Weinheim, 4th edition, p 66, 2003.
3. CE Schuster, *Handbook of Petrochemicals Production Processes*, (RA Meyers, ed.), McGraw-Hill, NY, p 14.45, 2005; MJ Kaus, 2005 *Petrochemical Seminar*, Mexico City (moved from Cancun), November 4, 2005.
4. M Mirra, *Handbook of Petrochemicals Production Processes*, (RA Meyers, ed.), McGraw-Hill, NY, p 14.59, 2005; see also A-A Finette and G ten Berge, *Handbook of*

Petrochemicals Production Processes, (RA Meyers, ed.), McGraw-Hill, NY, p 14.109, 2005.

5. J. P. Hogan, *J. Polymer Science, Part A-1*, 1970, 8, 2637.

6. J. P. Hogan, *Applied Industrial Catalysis*, 1983, 1, 149, Ch 6, BE Leach (editor), Academic, New York.

7. MP McDaniel, *Handbook of Heterogeneous Catalysis*, 2nd ed; 2007, chapter 15.1, G Ertl, H Knozinger, F Schuth, J Weitkamp, (editors), Wiley-VCH Verlag, Weinheim, Germany.

8. MP McDaniel, *Handbook of Heterogeneous Catalysis*, 1st ed, 1997, G Ertl, H Knozinger and J Weitkamp, (editors), VCH Verlagsgesellschaft, Weinheim, Vol 5, 2400.

9. E Benham and M McDaniel in *Kirk-Othmer Concise Encyclopedia of Chemical Technology*, 5th ed, A Seidel (editor), John Wiley & Sons, Inc., Hoboken, 2007, 590.

10. M. Smith, *Handbook of Petrochemicals Production Processes*, (RA Meyers, ed.), McGraw-Hill, NY, p 14.31, 2005

11. R Kuehl and G ten Berg, *Handbook of Petrochemicals Production Processes*, (RA Meyers, ed.), McGraw-Hill, NY, p 14.71, 2005

12. M deChellis and JG Griffin, US 5,352,749, April 26, 1993; see also FG Stakem, *International Conference on Polyolefins*, Society of Plastics Engineers, Houston, TX, February 22–25, 2009.

13. K Wiseman, *Handbook of Petrochemicals Production Processes*, (RA Meyers, ed.), McGraw-Hill, NY, p 14.131, 2005

14. T Korvenoja, H Andtsjo, K Nyfors and G Berggren, *Handbook of Petrochemicals Production Processes*, (RA Meyers, ed.), McGraw-Hill, NY, p 14.15, 2005.

15. ibid, p 14.23.

8

Downstream Aspects of Polyethylene

8.1 Introduction

In this chapter, subjects beyond polyethylene catalyst chemistry and manufacturing processes will be surveyed. After polyethylene is produced, it has only begun its journey into the intricate universe of applications and beyond. Usually, the raw polymer is formulated with additives, pelletized and transported to processors (who often add other additives). The processor melts the polymer and shapes it into myriad useful articles using a variety of fabrication techniques. During these processes, PE is exposed to conditions that can compromise its normally excellent chemical resistance and durability. The polymer is especially vulnerable during processing when it is heated to temperatures of 190 °C or higher and subjected to stress/strain forces that can cause cleavage of chemical bonds. These chemistries can affect polymer properties such as molecular weight and molecular weight distribution and render the polymer unsuitable for its intended use.

Polyethylene reaches the consumer in the form of a milk bottle or a bag for fresh vegetables in the supermarket or thousands of other goods that impact everyday life.

After the article's useful life is over, the question of how to dispose of the polymer must be addressed. Polyethylene is a durable material and is not readily biodegradable. This has become a contentious issue in some quarters. Laws have been enacted in a few communities banning certain polyethylene articles. A case study is the banning of plastic grocery sacks in San Francisco. The *raison d'etre* for the San Francisco ban included that the plastic sacks are "hard to recycle and easily blow into trees and waterways... They also occupy much-needed landfill space" (1). The ban requires San Franciscans to choose between biodegradable plastic bags or to bring their own reusable canvas bags. As of this writing, the cost of biodegradable bags is much higher (2x to 5x) relative to bags made from polyethylene. Whether such bans will solve the problem remains to be seen; that is, even biodegradable bags will be eyesores in the community until they biodegrade, which is a lengthy process under ambient conditions. (See section 8.5.) It has become commonplace to hear about local and national governments promulgating laws that require "cradle-to-grave" responsibility for manufacturers in the interest of "sustainable development."

Biomass-derived polymers are often touted as "green" alternatives to polyethylene and other plastics used for packaging. However, not all biopolymers are biodegradable (2). Moreover, as we shall see in section 8.5, degradability of biopolymers is sometimes overstated (3). In this chapter, we will quantify the contribution of plastics to municipal solid waste in the USA and examine some of the realities about biodegradability of "bioplastics."

These matters are far too broad and complex for a comprehensive discussion here. I will, however, attempt to provide perspectives on key downstream aspects, addressing such basic questions as:

- Why are additives necessary?
- What are the most common additives for polyethylene and how do they function?
- Why is rheology so important to the processing of polyethylene?
- What are the most important fabrication techniques for shaping polyethylene into consumer products?
- What companies are the major producers of polyethylene?
- What are the global volumes of the various types of polyethylene?
- What happens to polyethylene after its useful life is over?
- What are the biodegradable alternatives to polyethylene?
- What is the future of polyethylene?

References for more information about the topics discussed in this chapter are contained in each section.

8.2   Additives

Additives are essential to the performance of polyethylene. Indeed, modern polymers in general would not be viable without attributes that additives impart to the fully formulated resin. Additives are introduced

- to stabilize the polymer,
- to make the polymer easier to process, and/or
- to enhance its end use properties.

Many types of additives are used in polyethylene. A partial list is given below:

- antioxidants
- antistatic agents
- light (UV) stabilizers
- lubricants
- antimicrobials
- slip agents
- acid scavengers
- flame retardants
- polymer processing aids
- cross-linking agents
- anti-blocking agents
- colorants

Detailed discussions of the various types of additives used with polyethylene are not appropriate for an introductory text. However, there are many sources for specifics on additives to which the reader is referred (4–11). Excellent discussions are provided in the thorough handbooks edited by Zweifel (4, 5) and a recent text by Fink (6). A more compact overview of additives in film applications (a major end use of polyethylene) is provided by King (7). Specific topics are addressed in several recent trade magazine articles (8–11).

The optimal quantity of additive formulated into the resin varies with the type of polymer, the specific additive and its desired effect. For example, the amount of antiblock additives can be as high as several percent for some applications. ("Blocking" may be defined as the tendency of sheets of polyethylene film to "stick" together.) However, in most cases, the quantities of additives needed to achieve its function are in the range of 0.05 to 1% (500 to 10,000 ppm).

Polyethylene is susceptible to indirect attack by atmospheric oxygen. Free radicals are generated along the polymer chains by mechanisms that are still not fully understood (4). Radicals along the polymer chain react with atmospheric oxygen to generate peroxy radicals. This occurs throughout the life cycle of polyethylene. This is especially true of highly branched versions, such as LDPE, VLDPE and LLDPE, which have many tertiary carbon atoms (which are more susceptible than secondary or primary carbon atoms). Peroxy radicals undergo further reactions that can cause the polymer to degrade and lose mechanical strength. Hence, antioxidants that prevent or destroy free radicals are especially important to downstream applications of polyethylene.

Possible causes of radical formation in the polymer chain are residues of transition metals and peroxides used to catalyze polymerization and thermo-oxidative processes. Whatever causes homolytic bond scission to produce free radicals (eq 8.1 and 8.2), oxygen reacts with the resultant radical to generate hydroperoxide moieties. Key reactions involved in the initiation and propagation of radicals are illustrated in eq 8.1–8.4, where R_p is a polymeric fragment:

$$R_p\text{-}R_p \rightarrow 2\,R_p^{\cdot} \tag{8.1}$$

$$R_p\text{-}H \rightarrow R_p^{\cdot} + H^{\cdot} \tag{8.2}$$

$$R_p^{\cdot} + O_2 \rightarrow R_pOO^{\cdot} \tag{8.3}$$

$$R_pOO^{\cdot} + R_p\text{-}H \rightarrow R_pOOH + R_p^{\cdot} \tag{8.4}$$

Among the most important antioxidants are hindered phenols, which account for more than 50% of the total market for antioxidants sold into the global plastics industry as shown in Figure 8.1. Phosphite esters are second in importance as antioxidants for the global plastics industry. Polyethylene and polypropylene account for approximately ⅔ of the total market for antioxidants, as shown in Figure 8.2.

Antioxidants are classified as primary or secondary, depending upon how they react. Hindered phenols are primary antioxidants and function by donating a hydrogen to convert a peroxy radical to a hydroperoxide. Phosphites are among what are called secondary antioxidants and function as hydroperoxide decomposers. The ultimate outcome of these reactions is to convert the polymer bound radical to derivatives that are less destructive to the polymer.

A hindered phenol commonly used as an antioxidant is 2,6-di-*tert*-butyl-4-methylphenol (also known as butylated hydroxy toluene or "BHT"). Structures of BHT and other hindered phenol antioxidants are shown in Figure 8.3. Many of these complex structures have lengthy IUPAC names and are frequently called by trade names assigned by manufacturers, *e.g.*, Irganox® 1135 from Ciba (now BASF).

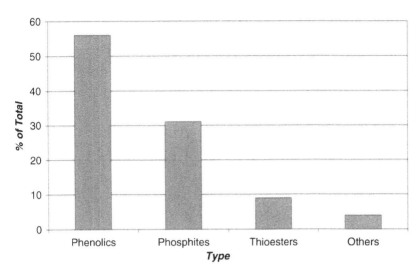

Figure 8.1 Global consumption of antioxidants in plastics.*

* Total consumption (1997): 207 million tons; "*Plastics Additives Handbook*," H. Zweifel, editor, Hanser Gardner Publications, Inc., Cincinnati, 5th Ed., 2001.

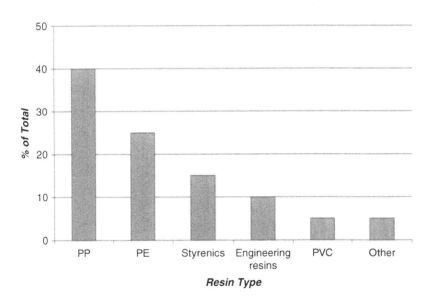

Figure 8.2 Global consumption of antioxidants by resin.*

Plastics Additives Handbook, H. Zweifel, editor, Hanser Gardner Publications, Inc., Cincinnati, 5th Ed., 2001, p 3.

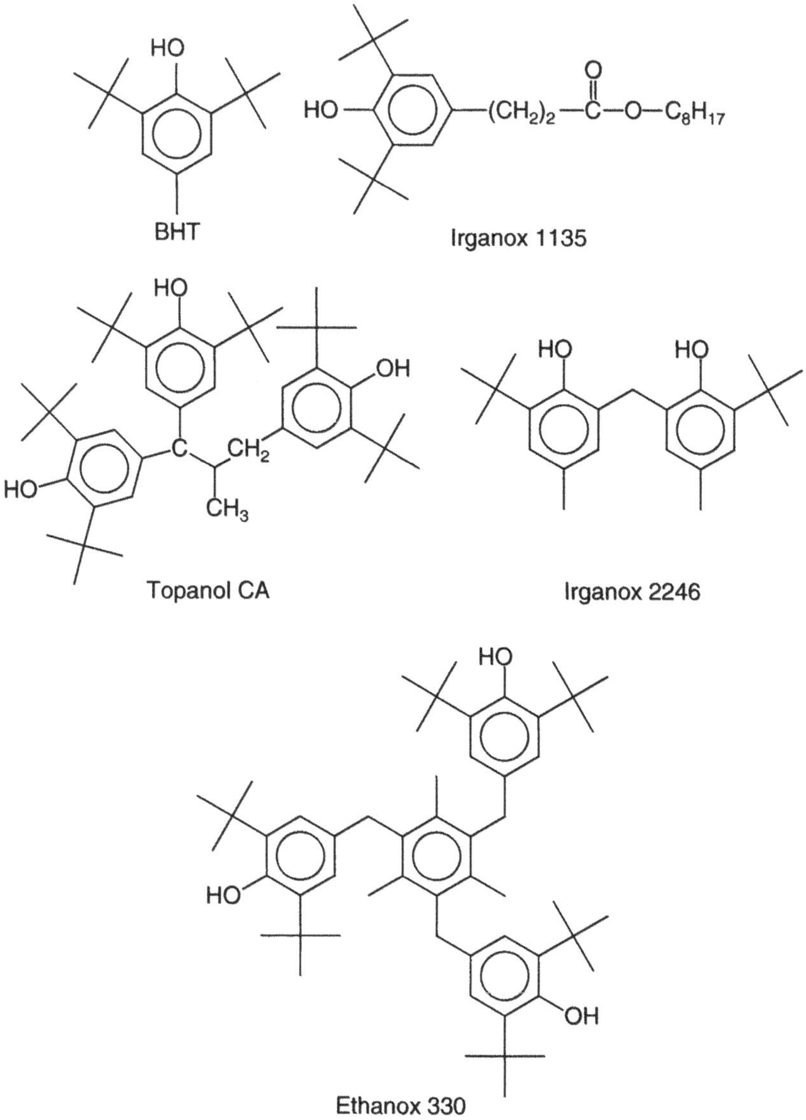

Figure 8.3 Structures of hindered phenol antioxidants. (Reproduced with permission from J. Fink, *A Concise Introduction to Additives for Thermoplastic Polymers*, Wiley-Scrivener Publishing, Salem, MA, 2010).

Simplified reactions that hindered phenols and phosphites undergo are shown in eq 8.5 and 8.6 below, where PhOH is a hindered phenol and PhO is a phenoxyl group:

$$PhOH + ROO \rightarrow PhO^{.} + ROOH \qquad (8.5)$$

$$(PhO)_3P + ROOH \rightarrow (PhO)_3P=O + ROH \qquad (8.6)$$

For example, if BHT were used in eq 8.5, the PhO˙ radical would be the structure below:

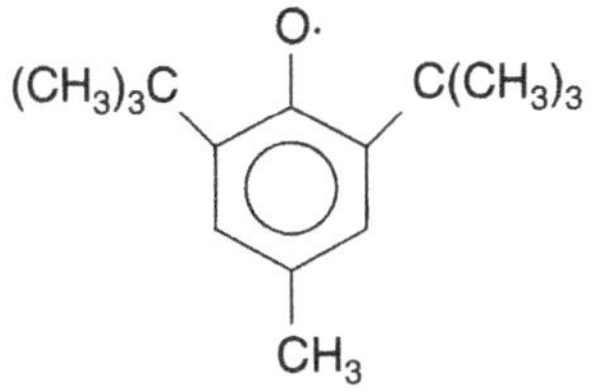

This structure has many resonance forms that result in a highly stabilized radical. In eq 8.6, a phosphite is oxidized to a phosphate and the hydroperoxide is reduced to an alcohol. Phosphites are often used in combination with hindered phenols. (Please see references 4–6 for details on functioning of antioxidants.)

8.3 Melt Processing

Rheology is defined as the science of the deformation and flow of matter. To enable polyethylene to be shaped into useful articles, the polymer must be melted and is typically heated to temperatures of ~190 °C. Even at such temperatures, the molten polymer is very viscous. Hence, rheological properties of molten polyethylene are crucial to its end use and much study has been devoted to this subject. Strict mathematical treatment of polymer rheology can become quite complex and is outside the scope of this text. However, general discussions of polymer rheology (12) and specifically for polyolefins (13–15) are available.

If a fluid (*e.g.*, water) flows in direct proportion to the force applied, it is said to exhibit "Newtonian" flow. However, the flow of molten polyethylene is not directly proportional to force applied, and polyethylene is said to exhibit "non-Newtonian" flow. Polyethylene becomes less viscous at higher stress ("shear"). This is called "shear thinning" and is typical of molten polyethylene.

As previously discussed in Chapter 1, melt index (MI) is a standard method for measuring flow of molten polyethylene and is indicative of molecular weight of the polymer. However, melt index is a limited indicator of rheological properties.

Techniques used for processing molten polyethylene are many. Some of the more important fabrication methods are listed below:

- extrusion (used to produce pipe, film, etc.)
- injection molding
- rotational molding
- blow molding
- compression molding (often used for UHMWPE)

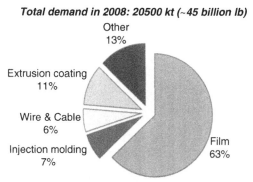

Figure 8.4 LDPE by processing method. (B. B. Singh, Chemical Marketing Resources, Webster, TX, March 26, 2007).

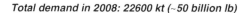
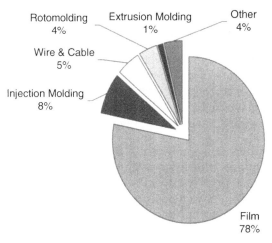

Figure 8.5 LLDPE by processing method. (B. B. Singh, Chemical Marketing Resources, Webster, TX, March 26, 2007).

The most suitable properties for polyethylene used in each fabrication method vary. For example, for blow molding applications, it is preferred to use HDPE having broad or bimodal molecular weight distribution. Figures 8.4–8.6 show how the major types of polyethylene are processed.

8.4   Markets

The global market in 2008 for all forms of polyethylene was estimated to be about 77 million metric tons (~169 billion pounds), with HDPE accounting

Total demand in 2008: 33500 kt (~74 billion lb)

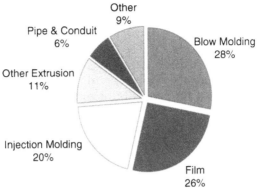

Figure 8.6 HDPE by processing method. (B. B. Singh, Chemical Marketing Resources, Webster, TX, March 26, 2007).

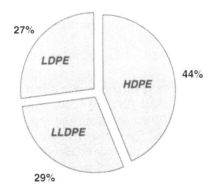

Total demand ~169 billion pounds (2008)

Figure 8.7 Global consumption of polyethylene by type (2008). (Total demand ~169 billion pounds (2008); C. Lee and B. B. Singh, Chemical Marketing Resources, Webster, TX, personal communication, June, 2009).

for about 44% of the total (16). The contribution of each major classification of polyethylene is shown in Figure 8.7. Overall growth of polyethylene is predicted to be about 5% per annum in the coming years. However, LDPE will grow more slowly (~2%). LLDPE and HDPE are expected to grow at about 6%. There are many caveats associated with these projections, including political unrest in major petroleum-producing regions and unstable economies. However, the projected overall growth is slightly lower than the actual overall growth rate (~6%) over the period 1990 to about 2002.

As shown in Figures 8.4 and 8.5, film applications are the most important end uses for LDPE and LLDPE, especially for food packaging (16). Blow molding and injection molding account for nearly half of all HDPE usage (see Figure 8.6). HDPE is also used in many film applications (17).

In general, LDPE provides film with better optical properties (*e.g.*, clarity and haze) and is easier to process. However, films made from LLDPE or HDPE display better mechanical properties (puncture resistance, tear strength, etc.), though they are more difficult to process. For this reason, LDPE is sometimes used as blend-stock with LLDPE and HDPE (18). The blended composition becomes easier to process while retaining good mechanicals.

As mentioned in Chapter 1, LLDPE is produced using α-olefins as comonomers. LLDPEs made with hexene-1 and octene-1 have better puncture resistance, impact strength and tear strength, but are more costly relative to LLDPE made with butene-1 comonomer.

In summary, mechanical strength of film made from the most common forms of low density polyethylene increases in the series:

LDPE < LLDPE (butene-1) < LLDPE (hexene-1) < LLDPE (octene-1)

Predictably, cost also generally increases going from left to right. The customer must then balance mechanical strength requirements for the specific application against material cost when selecting the type of polyethylene. When the mechanical strength required for the specific end-use can be met by LDPE (or LLDPE made with butene-1), there is no need to use LLDPE made with the more expensive comonomers hexene-1 or octene-1. Because LLDPE made with butene-1 combines good mechanical strength and low cost, butene-1 copolymer is the largest volume type of LLDPE. The breakdown of LLDPE by comonomer employed is shown in Figure 8.8. A comparison of film properties for LDPE, LLDPE made with butene-1 comonomer and LLDPE made with octene-1 comonomer is shown in Table 8.1.

LDPE has declined from about 35% of the total volume of all forms of polyethylene in 1990 to about 27% in 2008. This was caused by LLDPE (and to a lesser extent HDPE) displacing LDPE, mostly in film applications.

The largest global manufacturer of polyethylene in 2006 was The Dow Chemical Company followed by ExxonMobil. The top 10 industrial producers for 2006 are shown in Table 8.2, using the total of the three major types of polyethylene. In recent years such listings have been dynamic because of acquisitions, mergers and shifting trends in markets. For example, LyondellBasell (created by

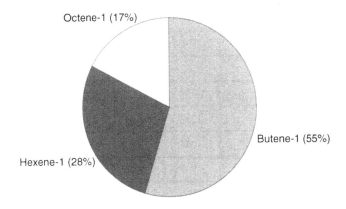

Figure 8.8 Global LLDPE by comonomer in 2008. (C. Lee, Chemical Marketing Resources, Inc., Webster TX, private communication June, 2009).

Table 8.1 Comparison of selected film properties of LDPE and LLDPE.*

Property	LDPE	LLDPE	LLDPE
Comonomer	none	butene-1	octene-1
Density (g/cc)	0.920	0.920	0.920
Melt index (g/10 min)	1.0	0.75	0.75

Elmendorf tear strength (g/mil)
  Machine direction
  Transverse direction
	170
	120
		65
		550
			400
			700

Impact strength (g/mil)	115	80	200
Puncture resistance (in-lb/mil)	3.0	6.0	8.0

*Dupont-Canada (now Nova) results on 1-mil film reported in Modern Plastics, p 127, April 1984.

the 2007 merger of Basell and Equistar, previously part of Lyondell) displaced SABIC as the third largest global producer of polyethylene, using data from Table 8.2. As this was being written, it was reported that that the Indian company Reliance Industries issued a "preliminary" bid to buy controlling interest in LyondellBasell (19). Clearly, rankings of top polyethylene producers will continue to fluctuate. In the coming years, Sinopec (the Chinese manufacturer) and companies with production capacities in the Middle East will continue to grow and will likely occupy the top rungs in Table 8.2.

Another complicating factor in estimating polyethylene volumes is the "swing" capability of some plants, *i.e.*, some reactors can be switched from LLDPE to HDPE (and vice versa) depending on market conditions.

Table 8.2 Top 10 global producers of polyethylene in 2006.*

Producer	HDPE	LDPE	LLDPE	Total
	(thousands of tons)			
Dow	1582	1687	4558	7827
ExxonMobil	2332	1596	2812	6740
SABIC	1068	751	2015	3834
Sinopec	769	982	1351	3102
INEOS	2203	300	520	3023
Chevron Phillips	2440	279	191	2910
Equistar (Lyondell)**	1386	665	513	2564
Basell***	1367	1045	0	2412
Borealis	697	730	764	2191
Total Petrochemical	1420	650	63	2133

* RJ Bauman, Nexant ChemSystems, International Conference on Polyolefins, Society of Plastics Engineers, Houston, TX, February 25–28, 2007.

** Lyondell merged with Basell in 2007

*** Now known as LyondellBasell.

8.5 Environmental

At this writing, economics are not favorable for recycle and/or reuse of polyethylene waste. The infrastructure to collect, separate and reprocess polyethylene is very limited. Consequently, at this point, most polyethylene waste goes into landfills. A common misconception by the public is that plastics are the major contributors to landfills. However, according to EPA figures, the reality is that plastics contributed only about 12% of total municipal waste in 2007 (see Figure 8.9), while paper products accounted for about 33%. The amount of plastics in landfills in 1970 was reported (20) to be about 11%, little different than in 2007. This was accomplished in part because items made from plastics with improved mechanical strength permitted "thinwalling" and "downgauging" over the decades, resulting in lower weights per unit (21).

The 12% figure for "plastics" in 2007 includes not only all forms of polyethylene but also other forms of thermoplastics such as PET, PP, PVC, polystyrene, etc. Though the actual figure is not readily available from the 2007 EPA statistics, a

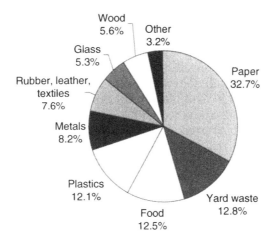

Figure 8.9 Municipal solid waste in the US in 2007. (Total 254 million tons, EPA figures).

Source: http://epa.gov/msw/facts.htm

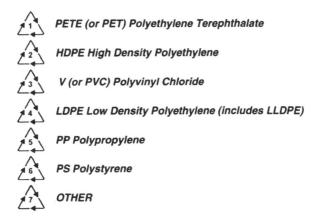

Figure 8.10 SPI coding of plastics.

plausible estimate of the contribution of polyethylene to municipal solid waste is probably less than 8%.

To aid recycling, the Society of the Plastics industry has issued numeric codes to identify the plastic used in fabricated articles. Each article should have an imprint of a triangle which encloses a number identifying the plastic used in its fabrication. LDPE (and LLDPE) are indicated by the number 4 and HDPE by the number 2. Codes for polyethylene and other plastics are shown in Figure 8.10.

At the supermarket checkout counter, customers are sometimes asked whether they prefer their groceries be put into "paper or plastic" bags. Those who choose paper because they think it is biodegradable, should consider the following: Though paper may be perceived by the public to be biodegradable, the reality is that paper in a landfill does not readily biodegrade. Studies have shown that newspapers buried in landfills are still readable after more than 15 years (20). Moreover, for those concerned about "global warming," paper products start with trees being harvested. Since removing trees from the environment results in more CO_2 in the atmosphere, the "carbon footprint" increases for those who choose paper. (After extensive reading on both sides of the issue, it is the author's humble opinion that global warming [or global cooling, for that matter] is largely a consequence of natural solar cycles. [See, for example, the discussion by Booker (21)]. The author further believes that mankind will not be able to change the trend significantly by regulating carbon emissions.)

One of the most visible environmental organizations is Greenpeace, known for their confrontational style over issues perceived to be important to the environment. Greenpeace has taken stances opposing certain types of chemicals, including those containing chlorine such as poly (vinyl chloride). Greenpeace has published a "Pyramid of Plastics" (Figure 8.11) that ranks plastics from least to most desirable from an environmental viewpoint. Because of its chlorine content, Greenpeace ranks PVC as the most objectionable plastic. Polyethylene ranks among plastics that are less objectionable (near the bottom of the pyramid). Of course, "biobased polymers" derived from renewable resources are most preferred by Greenpeace.

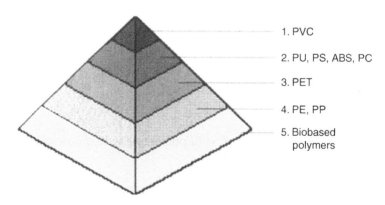

Figure 8.11 Greenpeace pyramid of plastics.

Source: http://archive.greenpeace.org/toxics/pvcdatabase/bad.html
Used with permission of Greenpeace.

Biopolymers and sustainable growth have been recently discussed by Singh (2). "Biopolymers" are defined as polymers made from "renewable" resources. However, not all biopolymers are biodegradable (2). For example, Dow and a Brazilian company called "Crystalsev" announced a joint venture in 2007 to manufacture LLDPE using ethylene produced from ethanol derived from sugarcane (22, 23). Though such LLDPE may be considered a biopolymer, it will not be any more biodegradable than LLDPE from petroleum-based ethylene.

One of the most highly developed biopolymers is poly (lactic acid) (PLA). In the USA, PLA is manufactured by NatureWorks at a plant in Nebraska using lactic acid derived from corn. (Lactic acid can also be obtained from other natural sources such as wheat or potatoes.) Poly (lactic acid) is produced by ring opening polymerization of the lactide, as shown in Figure 8.12.

PLA may be fabricated into film for packaging and is also made into fibers useful for carpeting (24). PLA is indeed biodegradable, but only under controlled composting conditions. Biodegradation of poly (lactic acid) requires temperatures of about 140 °F for many days to insure decomposition, ultimately into CO_2 and water. Unfortunately, such conditions are not typical of landfills (3) or of most backyard compost heaps. Hence, even plastic bags made from PLA will not quickly disappear from the natural environment. Anaerobic decomposition of PLA results in liberation of methane, an even more potent greenhouse gas than CO_2 (25–28).

A recent report by Singh (2) indicated that global production of poly (lactic acid) and other biopolymers is tiny compared to the quantity of polyolefins. Only

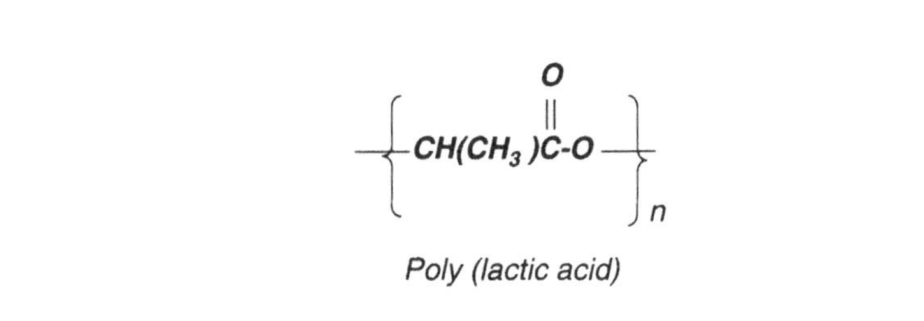

Figure 8.12 Poly (lactic acid). PLA usually obtained by ring opening polymerization of the lactide.

about 225,000 metric tons of biopolymers were produced in 2007, equivalent to about 0.25% of the global production of polyolefins. Singh reported that the cost of PLA is presently much higher than that of polyolefins and also suggested that for PLA to be truly competitive with polyethylene, the cost of lactic acid will need to be "on par with the price of ethylene"(2).

Clearly, bioplastics are a long way from becoming practical, cost-effective alternatives to polyethylene and other polyolefins. As the world seeks to attain sustained development, it is the author's opinion that polyethylene will become even more important. It is readily available from inexpensive raw materials by proven, efficient processes that produce very little waste during manufacture. Polyethylene exhibits versatility in applications and processing and lends itself to recyclability. Though there are unquestionably challenges ahead in creating the infrastructure for recycle and reuse, the future of polyethylene will remain bright well into the 21st century. It is certain to become a crucial part of the global effort to achieve economic growth while protecting the natural world.

References

1. C Goodyear, *San Francisco Chronicle*, March 28, 2007.
2. BB Singh, *International Conference on Polyolefins*, Society of Plastics Engineers, Houston, TX, February 22–25, 2009.
3. E Royte, *Smithsonian Magazine*, August, 2006; for entire article go to www.Smithsonian.com and search for "poly (lactic acid)."
4. H Zweifel (editor), *Plastics Additives Handbook*, Hanser Publishers, 5th edition, Munich, 2001
5. H Zweifel, R Maier and M Schiller (editors), *Plastics Additives Handbook*, Hanser Publishers, 6th edition, Munich, 2009.
6. J Fink, *A Concise Introduction to Additives for Thermoplastic Polymers*, Wiley-Scrivener Publishing, Salem, MA, 2010.
7. RE King, III, "Overview of Additives for Film Products", *TAPPI Polymer Laminations and Coatings Extrusion Manual*, T Butler (editor), TAPPI Press, September, 2000.
8. P Patel and B Puckerin, "A Review of Additives for Plastics: Colorants" *Plastics Engineering*, Society of Plastics Engineers, November, 2006.
9. P Patel and N Savargaonkar, "A Review of Additives for Plastics: Slips and Anti-blocks" *Plastics Engineering*, Society of Plastics Engineers, January 2007.
10. P Patel, "A Review of Additives for Plastics: Functional Film Additives," *Plastics Engineering*, Society of Plastics Engineers, August, 2007.
11. R Stewart, "Flame Retardants," *Plastics Engineering*, Society of Plastics Engineers, February 2009.
12. M Stevens, *Polymer Chemistry*, 3rd ed., Oxford University Press, New York, p 63, 1999.
13. A Peacock, *Handbook of Polyethylene*, Marcel Dekker, New York, p 220, 2000.
14. J White and D Choi, *Polyolefins*, Hanser Publishers, Munich, p 126, 2005.
15. C Vasile and M Pascu, *Practical Guide to Polyethylene*, Rapra Technology Ltd, p 97, 2005.

16. C Lee, Chemical Marketing Resources, Inc., Webster TX, private communication June, 2009.
17. B Singh, Chemical Marketing Resources, Inc., Webster TX, March 26, 2007.
18. MJ Kaus, *2005 Petrochemical Seminar*, Mexico City (moved from Cancun), November 4, 2005.
19. A Tullo, *Chemical & Engineering News*, p 10, November 30, 2009.
20. W Raftgey, *Saturday Night with Connie Chung*, AGS & R Communications, May 2, 1991.
21. C Booker, *The Real Global Warming Disaster*, Continuum, London, p 179, 2009.
22. Anon., *Chemical & Engineering News*, July 23, 2007, p 17.
23. AH Tullo, *Chemical &Engineering News*, September 29, 2008, p 21.
24. M McCoy, *Chemical & Engineering News*, December 14, 2009, p 7.
25. BS Mitra and R Gupta, *Global Warming and Other Eco-Myths*, (R. Bailey, editor), Competitive Enterprise Institute, p 145, 2002.
26. P Huber, *Hard Green*, Basic Books (Perseus Books Group), New York, p 63, 1999.
27. PJ Michaels and RC Balling, Jr, *Climate of Extremes*, Cato Institute, Washington, DC, p 14, 2009.
28. GA Olah, A Goeppert and GK Surya Prakash, *Beyond Oil and Gas: The Methanol Economy*, Wiley-VCH, Weinheim, p 41, 2006.

Glossary of Abbreviations, Acronyms and Terminology

Definitions of abbreviations, acronyms and terms are in context of polyethylene technology; may be different in other contexts

Abbreviation or Term	Definition
^{13}C	an isotope of carbon (as in carbon 13 NMR)
^1H	an isotope of hydrogen, a proton (as in proton NMR)
^{31}P	an isotope of phosphorus (as in phosphorus 31 NMR); used in Barron method
AA	acrylic acid
ABS	acrylonitrile-butadiene-styrene terpolymer
activator	another term for the metal alkyl cocatalyst in ZN catalyst systems
active aluminum	usually used to express "free TMAL" in methylaluminoxanes
aka	also knowm as
Al	aluminum
alkyl	generic name for hydrocarbyl groups (methyl, ethyl, isobutyl, *n*-butyl, etc.)
alpha olefins	linear 1-olefins; used as comonomers in LLDPE
amu	atomic mass units
anti-blocking agent	additive used to mimimize "blocking" (adhesion) of polyolefin films
antioxidant	additive used to minimize reaction of polyethylene with atmospheric oxygen
antistatic agent	additive used to minimize static electricity in polyolefin films
ARC	accelerating rate calorimetry
ASTM	American Society for Testing and Materials
atm	atmosphere

autoclave process	LDPE process wherein polymerization is conducted in a high pressure autoclave at very high T and P
BEM	n-butylethylmagnesium; also called MAGALA[a] BEM
BEM-B	n-butylethylmagnesium n-butoxide
blocking	tendency of PE films to cling (or "stick") together; problem solved by use of an anti-blocking agent
BOM	n-butyl-n-octylmagnesium
BOMAG[f]	see BOM
bp	boiling point
BR	butadiene rubber
BTSC	bis(triphenylsilyl) chromate
Bu	usually represents a normal butyl group ($CH_3CH_2CH_2CH_2-$); also abbreviated as n-Bu
BuCl or n-BuCl	n-butyl chloride (C_4H_9Cl)
C	Celsius (temperature measure, formerly "Centigrade")
catalyst	transition metal component of a ZN or single-site catalyst system
cc	cubic centimeter
CD	composition distribution
CGC	controlled geometry catalyst
cm^3	cubic centimeter
COC	cyclic olefin copolymer
cocatalyst	metal alkyl component of a ZN or single-site catalyst system
comonomer	a vinylic compound or simple olefin other than ethylene that is incorporated into the copolymer
conventional MAO	MAO produced by hydrolysis of TMAL
copolymer	produced by copolymerization of 2 vinylic compounds. e.g., ethylene and butene-1 or vinyl acetate
Cp	cyclopentadienyl (C_5H_5) group; often a ligand in metallocenes or SSCs
Cr	chromium (used in silica-supported Phillips catalysts for polyethylene)
CSTR	continuous stirred-tank reactor
CTA	chain transfer agents; used to control MW of polymers
d	density
DBM	dibutylmagnesium (commercial product is complex of DNBM and DSBM)
DEAC	diethylaluminum chloride
DEAI	diethylaluminum iodide
DEAL-E	diethylaluminum ethoxide
DEZ	diethylzinc
dg	decigram (0.1 g)
DIBAC	diisobutylaluminum chloride
DIBAL-H	diisobutylaluminum hydride
diene	type of monomer containing two olefinic sites, e.g., 1,3-butadiene; used in prod'n of elastomers

dimer	state of molecular association that involves two molecules per unit
DNBM[a]	di-n-butylmagnesium
DSBM	di-sec-butylmagnesium
DOT	Department of Transportation (regulates shipping containers, product classifications, etc.)
DP	degree of polymerization; number of repeating units (including end groups) in a polymer
DSC	differential scanning calorimetry
EAA	ethylene-acrylic acid copolymer
EASC	ethylaluminum sesquichloride; $((C_2H_5)_3Al_2Cl_3)$
EINECS	European Inventory of Existing Commercial Chemical Substances
elastomers	rubbery polymers; many made by copolymerization of olefins and/or dienes
EMA	ethylene-methacrylic acid copolymer
EPA	Environmental Protection Agency
EPDM	ethylene propylene diene monomer; rubbery copolymer produced with SSC and ZN catalysts
EPR	ethylene-propylene rubber
ESCR	environmental stress crack resistance
EU	European Union
EVA	ethylene-vinyl acetate copolymer
EVOH	ethylene copolymer with "hypothetical" comonomer vinyl alcohol
FAR	film appearance rating(s); from a standard test that measures defects in polyolefin films
FDA	Federal Drug Administration
Fe	iron (used in selected single-site catalysts)
fp	freezing point
free TMAL[c]	term applied to residual TMAL (or R_3Al) content of MAO and MMAOs
g	gram(s)
gas phase process	polymerization process wherein particles are suspended by circulating gas in a fluidized bed
GDP	gross domestic product
GPC	gel permeation chromatography; also called size exclusion chromatography (SEC)
Grignard reagent	RMgX (discovered by V. Grignard, usually in ether solution)
H	hydride ligand, as in diisobutylaluminum hydride (DIBAL-H)
H	hydrogen as in ROOH (hydroperoxides) or ROH (alcohols)
halide	generic designation for bromide, chloride or iodide; often represented in molecular formulas by X
HDPE	high density polyethylene (produced with ZN or Phillips catalysts)

HIC	household and industrial chemicals
HLMI	high load MI; determined under higher weight load (21.6 kg) than MI
HMW-HDPE	high molecular weight high density polyethylene
HMWPE	see HMW-HDPE
homopolymer	type of polymer produced with only one monomer, *i.e.*, without comonomer
IBAO	isobutylaluminoxane
ICI	Imperial Chemical Industries
in	inch
IPRA	abbreviation for product called "isoprenylaluminum" (see ISOPRENYL)
IR	infrared
isoprene	common name for 2-methyl-1,3-butadiene
ISOPRENYL	isoprenylaluminum (produced by reaction of TIBAL or DIBAL-H with isoprene)
IUPAC	International Union of Pure and Applied Chemistry
JV	joint venture
kg	kilogram
kt	kilotons (1 kt = 1000 metric tons = 2.2 million lb)
L	liter
LAO	linear alpha olefins (see alpha olefins)
lb	pound
LCB	long chain branching (usually in LDPE), *e.g.*, length of alkyl side chains
LDPE	low density polyethylene (produced with peroxides)
ligand	group or species bonded to a metal, *e.g.*, alkyl, alkoxide, hydride, chloride, etc
LLDPE	linear low density polyethylene (produced with ZN, SSC or Phillips catalysts)
m	meter
MA	methacrylic acid
MAGALA®	*mag*nesium-*al*uminum *a*lkyls; used as prefix for Akzo Nobel magnesium alkyls
MAO	methylaluminoxane
MD	machine direction; term used in polyolefin film testing (perpendicular to TD)
MDHDPE	see MDPE
MDPE	medium density high density polyethylene
Me	a methyl group; CH_3
metal alkyls	products containing at least one metal-carbon σ-bond
Met	abbreviation for metal in PE catalyst activity expressions
metallocene catalysts	type of single-site catalyst derived from π-bonded organometallic compounds

MFI	melt flow index; similar to MI but measured under different ASTM test and used mostly with PP
MFR	melt flow rate, another term for MFI; ASTM suggests MFR not be used for PE
Mg	magnesium
MgX_2	generic representation of a magnesium dihalide (such as magnesium chloride, $MgCl_2$)
MI	melt index; from an ASTM method; used as an indicator of MW of PE
micron (μ)	10^{-6} m
mil	10^{-3} in or ~25 microns
min	minute
MIR	melt index ratio (HLMI/MI); an indicator of the breadth of MWD of a polymer, used mostly with PE
mL	milliliter
mLLDPE	LLDPE produced with metallocene catalysts
mm	millimeter
MMAOs	generic term used by Akzo Nobel for modified methyla-luminoxanes; various types designated by suffix
M_n	number average MW
MPa	megapascal
mt	metric tons (1 mt = 2200 lb)
mVLDPE	VLDPE produced with metallocene catalysts
M_w	weight average MW
MWD	molecular weight distribution; key characteristic of polymers, also called polydispersity index
Ni	nickel (used in selected single-site catalysts)
NMR	nuclear magnetic resonance
organometallics	compounds that contain at least one metal-carbon bond; may be sigma or pi bond
Pd	palladium (used in selected single-site catalysts)
PDI	polydispersity index (M_w/M_n)
PET (or PETE)	poly(ethylene terephthalate)
Phillips catalyst	silica-supported chromium catalyst for HDPE developed by Phillips Petro. in the 1950s
PhO·	phenoxyl radical
PhOH	a hindered phenol (in context of discussion of antioxidants)
PLA	poly (lactic acid)
PMAO	polymethylaluminoxane; a less commonly used name for MAO
POE	polyolefin elastomer
polydispersity index	measure of MWD of a polymer; ratio of weight average MW to number average MW
polymer	a large molecule (or mixture of large molecules) consisting of repeating units of a monomer

polymerization	process whereby small molecules (monomers) are linked together to form large molecules
POP	polyolefin plastomer
PP	polypropylene (produced with ZN catalysts)
PPE	personal protective equipment
ppm	parts per million
PS	polystyrene, usually produced with peroxides
psd	particle size distribution
psig	pounds per square inch gauge
PU	polyurethane
PVC	poly(vinyl chloride)
pyridine titration	name given an analytical method for determining "free TMAL" in methylaluminoxanes
R	symbol for an alkyl group (methyl, ethyl, n-propyl, n-butyl, isobutyl, etc.)
ROH	an alcohol
ROOH	hydroperoxide
R_p	a polymeric (long chain) alkyl group
R_2Mg	generic representation of a dialkylmagnesium compound (such as DNBM)
R_3Al	generic representation of a trialkylaluminum compound (such as TMAL, TEAL, TIBAL, etc.)
replication	phenomenon whereby polymer particles assume the shape and psd of catalyst particles
rheology	study of the deformation and flow of fluids
RMgX	generic formula for an alkylmagnesium halide; see Grignard reagent
SABIC	Saudi Basic Industries Corporation
saponification	process in which an ester is converted to an alcohol and a carboxylic acid under basic conditions
SCB	short chain branching (usually in PE), $e.g.$, length of side chains in LLDPE
SEC	size exclusion chromatography (for determining MW and MWD of polymers);see also GPC
Si	silicon
silica	oxide of silicon (SiO_2); often used as support for PE catalysts
single-site catalysts	highly active transition metal catalysts[d]; many (not all) based on metallocenes
slurry process	process wherein polymerization is conducted in solvent in which polymer is insoluble
solution process	PE process wherein polymerization is conducted in "solution" at high temperature
SPE	Society of Plastics Engineers
SPI	The Society of the Plastics Industry, trade association established in 1937

SSC	single-site catalysts
stereochemical	chemistry dealing with the three-dimensional arrangement of atoms
suspension process	see slurry process
T	temperature
TBAO	tert-butylaluminoxane
TD	transverse direction; term used in polyolefin film testing (perpendicular to MD)
TEA	acronym for triethylaluminum
TEAL	acronym for triethylaluminum
TEB	triethylborane
terpolymer	copolymer in which three monomers are incoporated into the polymer
thermoplastic resins	polymers which can be melted repeatedly and formed into useful shapes
thermosetting resins	polymers which, once formed, cannot be melted and reshaped
Ti	titanium (most widely used metal in ZN catalysts; also used in single-site catalysts)
TIBAL	acronym for triisobutylaluminum
$TiCl_3$	titanium trichloride ("tickle 3") prepared by reduction of $TiCl_4$; early ZN catalyst; now largely obsolete
$TiCl_4$	titanium tetrachloride ("tickle 4"); raw material for many commercial ZN catalysts
TIPT	tetraisopropyl titanate
TMAL	acronym for trimethylaluminum
TNBAL	tri-n-butylaluminum
TNOAL	tri-n-octylaluminum
TREF	temperature rising elution fractionation
TSCA	toxic substance control act (part of EPA); all chemicals are to be listed with TSCA before mfg
tubular process	LDPE process wherein polymerization is conducted in a small diameter steel tube at very high T and P
UCC	Union Carbide Corporation
UHMWPE	ultrahigh molecular weight polyethylene
ULDPE	ultralow density polyethylene
Unipol	trademark for gas phase polymerization technology developed by Union Carbide (now Dow)
USA	United States of America
UV	ultraviolet
V	vanadium (used in ZN catalysts for polyethylene and synthetic rubber)
VA	vinyl acetate

VLDPE	very low density polyethylene
VOCl$_3$	vanadium oxytrichloride ("vocal 3"); raw material for ZN catalysts
XLPE	crosslinked polyethylene
Ziegler-Natta catalyst	combination of a metal alkyl and a transition metal[e] compound; used in olefin polymerizations
ZN catalyst	Ziegler-Natta catalyst
Zr	zirconium (widely used metal in single-site catalysts)

[a] MAGALA is used as a prefix for R$_2$Mg from Akzo Nobel. See also MAGALA.

[b] Registered trademark of Akzo Nobel.

[c] Also called "active aluminum."

[d] Most often involve Zr and Ti, though Fe, Co, Ni and Pd are also used. Allow extraordinary control of polymer molecular structure; usually activated by a methylaluminoxane.

[e] Commercial ZN catalysts mostly involve titanium compounds; relatively small amounts of vanadium are also used.

[f] Registered trademark of Chemtura (nee Crompton).

Trade Name Index

Index

Also of Interest

Check out these published and forthcoming related titles from Scrivener Publishing

A Concise Introduction to Additives for Thermoplastic Polymers by Johannes Karl Fink. Published 2010.
ISBN 978-0-470-60955-2.
Written in an accessible and practical style, the book focuses on additives for thermoplastic polymers and describes 21 of the most important and commonly used additives from Plasticizers and Fillers to Optical Brighteners and Anti-Microbial additives. It also includes chapters on safety and hazards, and prediction of service time models.

A Guide to Safe Material and Chemical Handling by Nicholas P. Cheremisinoff and Anton Davletshin. Published 2010.
ISBN 978-0-470-62582-8.
The volume provides an assembly of useful engineering and properties data on materials of selection for process equipment, and the chemical properties, including toxicity of industrial solvents and chemicals.

Handbook of Engineering and Specialty Thermoplastics

Volume One: Polyolefins and Styrenics by Johannes Karl Fink. Published 2010. ISBN 978-0-470-62483-5.

Volume Two: Polyethers and Polyesters edited by Sabu Thomas and Visakh P.M. Forthcoming late 2010.

Volume Three: Nylons edited by Sabu Thomas and Visakh P.M. Forthcoming late 2010.

Volume Four: Water Soluble Polymers edited by Johannes Karl Fink. Forthcoming 2011.

Miniemulsion Polymerization Technology edited by Vikas Mittal. Forthcoming summer 2010.

The book is a ready reference for the background information as well as advanced knowledge regarding the applications of miniemulsion polymerization technology.

Printed and bound by CPI Group (UK) Ltd, Croydon, CR0 4YY

16/04/2025

14658453-0004